4th International Conference on Emerging Technologies in Computer Engineering: Data Science & Blockchain Technology (ICETCE 2021)

AA007239

IOP Conference Series: Materials Science and Engineering
Volume 1131

Jaipur, India
3 - 4 February 2021

Editors:

Anil Chaudhary
C.M. Choudhary
Pankaj Dadheech
Mehul Mahrishi
Linesh Raja
Ankit Kumar

ISBN: 978-1-7138-3409-0
ISSN: 1757-8981

Printed from e-media with permission by:

Curran Associates, Inc.
57 Morehouse Lane
Red Hook, NY 12571

Some format issues inherent in the e-media version may also appear in this print version.

This work is licensed under a Creative Commons Attribution 3.0 International Licence.
Licence details: http://creativecommons.org/licenses/by/3.0/.

No changes have been made to the content of these proceedings. There may be changes to pagination and minor adjustments for aesthetics.

Printed with permission by Curran Associates, Inc. (2021)

For permission requests, please contact the Institute of Physics at the address below.

Institute of Physics
Dirac House, Temple Back
Bristol BS1 6BE UK

Phone: 44 1 17 929 7481
Fax: 44 1 17 920 0979

techtracking@iop.org

Additional copies of this publication are available from:

Curran Associates, Inc.
57 Morehouse Lane
Red Hook, NY 12571 USA
Phone: 845-758-0400
Fax: 845-758-2633
Email: curran@proceedings.com
Web: www.proceedings.com

TABLE OF CONTENTS

Preface

The **4th International Conference on Emerging Technologies in Computer Engineering: Data Science & Blockchain (ICETCE-2021)** was held at Swami Keshvanand Institute of Technology, Management and Gramothan (SKIT), Jaipur, Rajasthan, India, on February 3rd – 4th, 2021 with the aim to provide a platform for researchers working in the field of computer science to present the state-of-the- art methodologies, novel ideas, and high-quality research papers.

The inaugural session of the conference witnessed plenary talks by **Dr. Arun K Somani**, Associate Dean of Research, College of engineering, Iowa State University and **Prof. Seeram Ramakrishnan**, Vice President- Research Strategy & Professor – Faculty of Engineering, National University of Singapore (NUS) Singapore. The conference guests were **Ms. Mona Bhardwaj**, Country Manager, University Relations, IBM Research, India, and **Mr. Victor Sundarraj A.**, Head – Engineering and Emerging Technologies Academy, Infosys Ltd.

Numerous participants and representatives from Academia, Industry, Media, Government of Rajasthan, Research and Student community participated in the conference. A total of 110 research papers were submitted from across the world. All papers undergone rigorous peer-review process and finally 22 papers were presented and selected for inclusion in **IOP Conference Series: Materials Science & Engineering**.

Through this forum we would like to take a chance to thank people behind the success of this conference. First and foremost, we want to thank IOP Science and their representatives and Conference Quality Specialist for their continuous support and regular reminders if somehow, we got off track. We wish to thank the management of Swami Keshvanand Institute of Technology, Management and Gramothan **(SKIT),** Jaipur, Rajasthan, India, for providing the best infrastructure to organize this conference. We are also thankful to our technical partners **IBM India, Infosys Ltd. and Natural group** for their aggressive involvement and support during such technical events. The editors would like to acknowledge the hard work of all the authors involved in this project and of the peer reviewer who took part in the review process.

We hope that this conference proceeding will prove to be useful.

Editors:

- Prof. (Dr.) Anil Chaudhary, HOD, Department of Information Technology, Swami Keshvanand Institute of Technology, Management & Gramothan (SKIT), Ramnagaria, Jagatpura, Jaipur, Rajasthan, India-302017, Email ID: hodit@skit.ac.in

Content from this work may be used under the terms of the Creative Commons Attribution 3.0 licence. Any further distribution of this work must maintain attribution to the author(s) and the title of the work, journal citation and DOI.
Published under licence by IOP Publishing Ltd

IOP Conf. Series: Materials Science and Engineering 1131 (2021) 011001 doi:10.1088/1757-899X/1131/1/011001

- Prof.(Dr.) C. M. Choudhary, Professor, Department of Computer Science & Engineering, Swami Keshvanand Institute of Technology, Management & Gramothan (SKIT), Ramnagaria, Jagatpura, Jaipur, Rajasthan, India-302017, Email ID: chothmal@skit.ac.in
- Dr. Pankaj Dadheech, Associate Professor, Department of Computer Science & Engineering, Swami Keshvanand Institute of Technology, Management & Gramothan (SKIT), Ramnagaria, Jagatpura, Jaipur, Rajasthan, India-302017, Email ID: pankaj@skit.ac.in
- Mehul Mahrishi, Associate Professor, Department of Information Technology, Swami Keshvanand Institute of Technology, Management & Gramothan (SKIT), Ramnagaria, Jagatpura, Jaipur, Rajasthan, India-302017, Email ID: mehul@skit.ac.in
- Dr. Linesh Raja, Assistant Professor, Manipal University Jaipur, Rajasthan, India, Email ID: lineshraja@gmail.com
- Ankit Kumar, Assistant Professor, Department of Computer Science & Engineering, Swami Keshvanand Institute of Technology, Management & Gramothan (SKIT), Ramnagaria, Jagatpura, Jaipur, Rajasthan, India-302017, Email ID: iiita.ankit@gmail.com

List of Committees (ICETCE-2021)

Name of the Conference:

4th International Conference on "Emerging Technologies in Computer Engineering: Data Science & Blockchain Technology (ICETCE-2021)" will be organized on 3rd – 4th February 2021.

Venue (Institute / Hotel / Convention Center):

Department of Computer Science & Engineering and Information Technology (Both are NBA accredited),

Swami Keshvanand Institute of Technology, Management & Gramothan (SKIT), Ramnagaria, Jagatpura, Jaipur-302017 (Rajasthan, India)

Conference website:

https://icetce.skit.ac.in/index

General Chair:

- *Prof. Seeram Ramakrishna, Vice President Research Strategy, Professor, Faculty of Engineering National University of Singapore (NUS), Singapore. Email: : seeram@nus.edu.sg*

H-index= 139, i10-index 770, Citations= 85700

https://scholar.google.co.in/citations?user=a49NVmkAAAAJ&hl=en

Webpage: http://me.nus.edu.sg/staff.php?id=2060

For authentic h-index of Prof. Seeram Ramakrishna:- (http://www.webometrics.info/en/node/58)

Conference Chair:

- *Prof. (Dr.) Arun K. Somani, Associate Dean for Research, College of Engineering, Iowa State University, Ames, USA*

  *H-index=48, i10-index=171, Citations=*9215

  https://scholar.google.co.in/citations?user=SYZ5R-8AAAAJ&hl=en

  Webpage: http://class.ece.iastate.edu/arun/

  dblp Link: https://dblp.uni-trier.de/pers/hd/s/Somani:Arun_K= (222 records)

Technical Program Chair:

Prof. (Dr.) Arun K. Somani, Associate Dean for Research, College of Engineering, Iowa State University, Ames, USA

Dr. G. R. Sinha, Professor, Myanmar Institute of Information Technology (MIIT) Mandalay, Myanmar

Dr. Basant Agarwal, Assistant Professor, Department of Computer Science & Engineering, Indian Institute of Information Technology Kota (IIIT Kota), India

Dr. Pankaj Dadheech, Associate Professor & Dy. HOD, Department of Computer Science & Engineering, SKIT, India

Dr. Tarun Kumar Sharma, Professor and Dean, Department of Computer Science & Engineering, School of Engineering and Technology, Shobhit University Gangoh, Saharanpur.

Organizing Chair:

- Prof. Anil Chaudhary, *IEEE Senior Member,* Professor & Head, Department of Information Technology, SKIT, *Jaipur.*

  H-index=6, i10-index=2, Citations=108
  https://scholar.google.co.in/citations?user=UoU64kkAAAAJ&hl=en&oi=ao

Telephone: 0141-5160400, 2752165, 2752167 (Ext. 286),

Mobile No: 9829133780, E-mail Id: hodit@skit.ac.in

IOP Conf. Series: Materials Science and Engineering 1131 (2021) 011001 doi:10.1088/1757-899X/1131/1/011001

- *Prof. (Dr.) C. M. Choudhary,*

  Fellow Institution of Engineers (FIE), Professor and Head, Department of Computer Science & Engineering, SKIT, Jaipur.

  Telephone: 0141-5160400, 2752165, 2752167 (Ext. 283),

  Mobile No: 9414305959, E-mail Id: hodcs@skit.ac.in

Publication Chair:

Mr. Mehul Mahrishi, Associate Professor & Dy. HOD, IT Department, SKIT

Technical Programme Committee (TPC)

- *Prof. Sandeep Sancheti, Vice Chancellor, SRM, Chennai*
- *Dr. Virendra Singh, Associate Professor, Department of Electrical Engineering, Indian Institute of Technology (IIT) Bombay, India*
- *Prof. G. R. Sinha, Adjunct Professor, International Institute of Information Technology (IIIT), Bangalore & IEEE Executive Council Member, MP Subsection, India*
- *Prof. R. K. Joshi, Department of Computer Science & Engineering, IIT Bombay, India.*
- *Dr. Mitesh Khapara, Department of Computer Science & Engineering, IIT Madras, India*
- *Prof. R S Shekhawat, Professor & Director- School of Computing & IT, Manipal University, India*
- *Prof. Y.K. Vijay, President, VGU, Jaipur*
- *Dr. Mahesh Chandra Govil, Director, NIT Sikkim, India*
- *Dr. Manoj Singh Gaur, Director, Indian Institute of Technology, Jammu, India*
- *Dr. Virendra Singh, Professor, Department of Computer Science & Engineering, IIT Bombay, India*
- *Dr. Vijay Laxmi, Professor, Department of Computer Engineering, MNIT, Jaipur*
- *Dr. Neeta Nain, Associate Professor, Department of Computer Engineering, MNIT, Jaipur*
- *Dr. Namita Mittal, Associate Professor, Department of Computer Engineering, MNIT, Jaipur*
- *Dr. Emmanuel Shubhakar Pilli, Associate Professor, Department of Computer Engineering, MNIT, Jaipur*
- *Dr. Mushtaq Ahmed, Associate Professor, Department of Computer Engineering, MNIT, Jaipur*
- *Dr. Yogesh Meena, Associate Professor, Department of Computer Engineering, MNIT, Jaipur*
- *Dr. S.C. Jain, Professor, Department of Computer Engineering, RTU, Kota*
- *Dr. Dayanand Kumar, Principal Integration Engineer, Micron Technology, Taiwan*
- *Dr. Mohammad Shariq, Professor, Jazan University, Saudi Arabia*
- *Prof. Mohammd Salim, MNIT Jaipur, India*
- *Prof. Ghanshyam Singh, MNIT Jaipur, India*
- *Dr. Vipin Pal, NIT Meghalaya, India*
- *Dr. Yogita, NIT Meghalaya, India*
- *Dr. Sanjeev Kumar Metya, NIT Arunachal Pradesh, India*
- *Prof. Lakshmi Prayaga, University of West Florida, Pensacola, FL*
- *Prof. Ehsan O. Sheybani, Virginia State University, USA*
- *Prof. Sanjay Kaul, Fitchburg State University, USA*
- *Dr. Abhimanyu Singh Garhwal, Massey University, Auckland, New Zealand.*
- *Dr. Mani Madhukar, Program Manager, University Relations, IBM India Pvt. Ltd.*
- *Dr. Manish Pokharel, Department of Computer Science & Engineering, Kathmandu University, Nepal*

IOP Conf. Series: Materials Science and Engineering 1131 (2021) 011001 doi:10.1088/1757-899X/1131/1/011001

- *Dr. Nistha Keshwani, Department of Computer Science, Central University of Rajasthan, India*
- *Dr. Shashnak Gupta, Department of Computer Science, BITS Pilani, India*
- *Dr. Rajiv Ratn, Department of Computer Science, IIIT Delhi, India*
- *Dr. Vinod P Nair, Department of Mathematics, University of Padua, Italy*
- *Dr. Adrian Will, National Technological University, Tucumán, Argentina*
- *Dr. Chhagan Lal, Department of Mathematics, University of Padua, Italy.*
- *Dr. Subrata Mukhopadhyay, Member (ex-officio), IEEE Delhi Section*
- *Prof. K. Subramanian, Chair, IEEE Delhi Section*
- *Dr. A. Murali M. Rao, Past Chair, IEEE CS, Delhi Section*
- *Dr. Nisheeth Joshi, Department of Computer Science, Banasthali University, India*
- *Dr. Maadvi Sinha, Department of Computer Science & Engineering, Birla Institute of Technology, India.*
- *Mr. Daman Dev Sood, Member, IEEE Delhi Section, India*
- *Dr. Soujanya Poria, Nanyang Technological University, Singapore*
- *Dr. Vaibhav Katewa, University of California, Riverside California, USA*
- *Dr. Sugam Sharma, Iowa State University, USA*
- *Dr. Xiao-Zhi Gao, Lappeenranta University of Technology, Finland*
- *Dr. Deepak Garg, Professor and Head, CSE at Bennett University, Greater Noida*
- *Dr. Pranav Dass, Galgotias University, Uttar Pradesh, India*
- *Dr. Linesh Raja, Amity University Rajasthan, India*
- *Dr. Vijander Singh, Amity University Rajasthan, India*
- *Dr. Gai-Ge Wang, Professor, Jiangsu Normal University, China*
- *Dr. Piyush Maheshwari, Amity University, Dubai, UAE*
- *Dr. Wenyi Zhang, Professor, University of Michigan, USA*
- *Dr. Tapas Badal, Associate Professor, Department of Computer Science and Engineering, SKIT, India*
- *Dr. Janos Arpad Kosa, Kecskemet College, Hungary*
- *Dr. Dongxiao He, Tianjin University, China*
- *Dr. Thoudam Doren Singh, Department of Computer Science & Engineering, Indian Institute of Information Technology Manipur*
- *Dr. Dharm Singh Jat, Namibia University of Science and Technology, Namibia, Africa*
- *Dr. Vishal Goyal, Associate Professor, Department of Computer Science, Punjabi University, India*
- *Dr Dushyant Singh, Department of Computer Science and Engineering, MNNIT Allahabad, India*
- *Mr. Amit Kumar Gupta, Scientist 'D', DRDO, Hyderabad*
- *Pallavi Kaliyar, Department of Mathematics, University of Padua, Italy*
- *Dr. Sumit Srivastava, Senior Member IEEE Delhi Section & Professor (CSE), Manipal University.*
- *Mr. Ripudaman, Director, Natural Group*
- *Mr. Sumit Srivastava, Consultant, Pratham Software Pvt. Ltd., Jaipur*
- *Mr. Nitin Purohit, Wollo University, Dessie, Ethiopia*
- *Mr. Ankush Vasthistha, Department of Computer Science, NUS, Singapore*
- *Prof. (Dr.) Dinesh Goyal, Professor & Dean (Academics), SGVU*
- *Prof. (Dr.) Reena Dadhich, HOD, Department of Computer Science & Informatics, University of Kota, Rajasthan*
- *Dr. Arvind K. Sharma, Assistant Professor (CS), University of Kota, Rajasthan*
- *Dr. Ramesh C. Poonia, Associate Professor (CS), AIIT, Amity University, Rajasthan*
- *Dr. Vijendra Singh, North Cap University, India*
- *Dr. Brijesh Kumbhani, IIT Ropar, India*
- *Dr. Shashi Kant Sharma, IIIT Ranchi, India*
- *Dr. Nikhil Deep Gupta, NIT Nagpur, India*
- *Dr. Rimpy Bishnoi, LNMIIT, Jaipur, India*

IOP Conf. Series: Materials Science and Engineering 1131 (2021) 011001 doi:10.1088/1757-899X/1131/1/011001

- *Mr. Ashok Chauhan, Senior Scientist, CEERI Pilani, India*
- *Dr.Ing Nidhi Chaturvedi, Principal Scientist, CEERI Pilani, India*
- *Dr. Kunwar Pal, NIT Sikkim, India*
- *Mr. Sandeep Saini, Myanmar Institute of Technology, Myanmar*
- *Dr. C. Periasamy, MNIT Jaipur, India*
- *Mr. Harish Indouria, Skillrock Technologies, India*
- *Mr. Ashish Sharma, IIIT Kota, India*
- *Dr. Rajat Goel, SKIT Jaipur, India*
- *Dr. Niketa Sharma, SKIT Jaipur, India*
- *Mr. Yogendra Gupta, SKIT Jaipur, India*
- *Mr. Maninder Singh Nehra, Govt. Engineering College Bikaner, India*
- *Dr. Sunita Gupta, Associate Professor, SKIT Jaipur. India*
- *Dr. Rakesh K. Bhujade, Associate Professor, Technocrats Group of Institution, Bhopal MP, India*
- *Mr. Mahesh Pareek, Programming Officer, ONGC, New Delhi Area, India*
- *Ms. Kamaljeet Kaur, Campus Connect and Agile, ETA, Chandigarh DC, Infosys Ltd.*
- *Mr. Rupesh Jain, Practice Manager, Wipro Technologies Ltd.*
- *Mr. Amit Kumar Gupta, Scientist 'D', DRDO, Hyderabad*
- *Dr. Gaurav Singhal, Assistant Professor, Bennett University, Greater Noida*
- *Dr. Lokesh Sharma, Assistant Professor, Galgotia University, Greater Noida*
- *Dr. Ankit Vidhyarthi, Department of Computer Science and Engineering, Bennett University, Greater Noida*
- *Dr. Vikas Tripathi, Assistant Professor, Graphic Era University, Dehradun*
- *Dr. Prakash Choudhary, Department of Computer Science and Engineering, NIT Manipur, Manipur, India*
- *Dr. Smita Naval, Department of Computer Science, IIIT Kota*
- *Dr. Subhash Panwar, Government Engineering College, Bikaner, India*
- *Ms. Richa Jain, Assistant Professor, Banasthali University, Newai, Tonk*
- *Mr. Pankaj Jain, Assistant Professor, K.N. Modi University, Newai, Tonk*
- *Dr. Manoj Bohra, Associate Professor, Manipal University, India*
- *Dr. Rajbir Kaur, Associate Professor, The LNM Institute of Information Technology*
- *Dr. Poonam Gera, Associate Professor, The LNM Institute of Information Technology*
- *Sh. Madan Mohan Agarwal, Associate Professor, Birla Institute of Technology*
- *Dr. Ajay Khunteta, Professor, Poornima College of Engineering, Jaipur*
- *Sh. Subhash Gupta, Associate Professor, Birla Institute of Technology*
- *Mr. Devendra Gupta, Lecturer, Govt. Polytechnic College Tonk*
- *Dr. Baldev Singh, Professor, Vivekanand Institute of Technology*
- *Dr. Kavita Choudhary, Associate Professor, J.K. Laxmipat University*
- *Dr. Vikas Tripathi, Assistant Professor, Graphic Era University, Dehradun*
- *Dr. Rahul Dubey, Assistant Professor, Madhav Institute of Technology and Science, Gwalior*
- *Dr. Smita Agarwal, Associate Professor, Global Institute of Technology*
- *Dr. Sonal Jain Associate Professor, Department of Computer Science, JK Lakshmipat University*
- *Mr. Shrawan Ram, Associate Professor, MBM Engineering College, Jodhpur*
- *Dr. Sonali Vyas, Assistant Professor, AIIT, Amity University, Rajasthan*

Finance Chair / Treasurer

• *Mr. Vipin Jain, Associate Professor, IT Department, SKIT*

• *Mr. Kailash Soni, Assistant Professor, CSE Department, SKIT*

IOP Conf. Series: Materials Science and Engineering 1131 (2021) 011002 doi:10.1088/1757-899X/1131/1/011002

Peer review declaration

All papers published in this volume of *IOP Conference Series: Materials Science and Engineering* have been peer reviewed through processes administered by the Editors. Reviews were conducted by expert referees to the professional and scientific standards expected of a proceedings journal published by IOP Publishing.

- **Type of peer review: Double-blind**
- **Conference submission management system: EasyChair**
- **Number of submissions received: 110**
- **Number of submissions sent for review: 76**
- **Number of submissions accepted: 28**
- **Acceptance Rate (Number of Submissions Accepted / Number of Submissions Received X 100): 25.45%**
- **Average number of reviews per paper: 3**
- **Total number of reviewers involved: 260**
- **Any additional info on review process**
    - *The paper submission shall be through Easy Chair.*
    - *At least three independent reviewers will review the paper.*
    - *The review will be a double blind peer review process.*
    - *A maximum of 2-3 papers would be assigned to a single reviewer.*
    - *Based on the reviews received, the technical program chair will send the final notification of the acceptance /rejection of the paper.*
    - *In case of any contradictory views of reviewers, The TPC chair would invite one more review of the paper to take the decision.*
    - *Plagiarism will be checked through Turnitin plagiarism software.*

The Technical Program Committee (TPC) is responsible for the peer review process.

The committee is listed on the website of the conference.

- **Contact person for queries: Dr. Anil Chaudhary (hodit@skit.ac.in)**

Content from this work may be used under the terms of the Creative Commons Attribution 3.0 licence. Any further distribution of this work must maintain attribution to the author(s) and the title of the work, journal citation and DOI.
Published under licence by IOP Publishing Ltd

Index Point Detection for Text Summarization using Cosine Similarity in Educational Videos

Rekha Sharma, Mehul Mahrishi, Sudha Morwal and Girish Sharma
KanoriaPGMahilaMahavidyalaya, Jaipur, Rajasthan, India[1]

SwamiKeshvanandInstituteofTechnology, Management&Gramothan, Jaipur, Rajasthan, India[2]
BanasthaliVidhyapith, Niwai, Rajasthan, India[3]

SwamiKeshvanandInstituteofTechnology, Management&Gramothan, Jaipur, Rajasthan, India[4]

E-mail: rekha04sharma@gmail.com

Abstract. Explosive growth in digital content creates a trend of video-based learning and knowledge sharing. Several educational videos are shared by various trainers/ bloggers/ instructors daily. It is now evident that even during emergencies like the recent ongoing CoVid-19 outbreak, video lectures are the saviors when the whole world comes to rest. This rapid and expansive growth has encouraged the researchers to efficiently and automatically index, browse, and retrieve video data. This research aims to process videos and precisely identify the index points which are discussed within the video. Experimental results show a good performance of the topic detection model in finding keywords and reducing dimensionality.
Keywords: Cosine Similarity, Key-frame Identification, Optical Character Recognition,Text Extraction ,Text Recognition,Video Frames

1. Introduction

Segmentation splits a video stream into different camera-takes/scenes or shots.These are nothing but the correlated sets of a contiguous picture frame sequence taken by a camera from the moment it begins to capturing the moment it finishes. [1][2]. Content-wise, shots are homogeneous and possess a degree of visual uniformity. The idea is to extract the video content and analyze it for information and knowledge retrieval. This paper intends to offer a framework that assists in the video summarization technique by extracting keywords as and when they occur in the video. Shot Boundary Detection is done through key-frame identification using Cosine Similarity with a static threshold value.

1.1. Textual content in a Video

Text in a video sequence can exists in one of the two forms viz.

- Scene Text: It is also called graphic text and is displayed in a video sequence during the video making, for example, text written on boards, banners, and hoardings.
- Artificial text: is the text that is introduced explicitly into the video through editing. It is always displayed at a specific position in the video. Examples are names, data about the video, or interpretation of discoursed in video film, captions in a video.

Content from this work may be used under the terms of the Creative Commons Attribution 3.0 licence. Any further distribution of this work must maintain attribution to the author(s) and the title of the work, journal citation and DOI.
Published under licence by IOP Publishing Ltd

1.2. Online Video Lectures
Recording video lectures and presentations for enhancing the teaching-learning mechanism is now longstanding in India.[3] [4]. With the support of the Ministry of Human Resource Development (MHRD), institutes of national importance in India started recording the video lectures, which substantially increased the self-paced learning for students and academicians. This repository is now huge and authentic that many articles have been published, and much research is possible for data scientists by using this data-set [5]. This data-set is very useful in many research domains like natural language processing, speech recognition, image, and video analysis.

2. Related Work
Text summarization is the process that automatically gives the precise meaning of the given document and attracting the attention of Natural Language Processing researchers.[1] Proposed an algorithm in which the OpenCV library is used to decode video into frames. [6] proposed that Clustering and dictionary learning together can be employed to determine keyframes, and additional local and global features of these frames can is determined. Color features [7],[8], moments of inertia [9] as well as local binary features[10] are most predominantly used for representing image frames in a video. Table 1 represents the literature regarding text and feature extraction techniques used by different authors in their state-of-the-art research.
Few methods came into existence that uses divide-and-conquer algorithms to segment videos and determines keyframes [11]. [12]Using an intra-modality and inter-modality fusion mechanism to select sub-summaries, the aggregated attention curve was created and all sub-summaries were finally combined into an aggregated summary. [13]proposed a dual-threshold method for video segmentation. Different keyframes are extracted from each segment. Scale-Invariant Feature Transform (SIFT) features are extracted from the keyframes of the segments. It is proposed in SVD-based method (Singular Value Decomposition) ,two video frames with SIFT point set descriptors are suggested to match.
[14] figures out the mutual information method according to information theory and confirms that it improves the extraction efficiency.

3. Index Point Detection Framework
This research discovered that the frequency of words, frequency of n-grams, and several first-time words in a video provide valuable information for video segmentation by topic. The research is intended to identify semantic and topic-based index points in educational videos so that the video has discussed topics that can be quickly and efficiently listed. These index points are the topic/subtopic discussed in the video lecture. Data sets of online video lectures, tutorials, open course-ware, and similar resources with variable playback time and specifically the videos that contain lecture slides are used as input. As represented in figure 1, the proposed framework is robust and exhibits good performance for video text detection and recognition.

For similarity measures, cosine similarity is used, and we adopted a threshold value while measuring the similarity among video frames. Videos with most graphs, statistical data, and diagrams are avoided in the data set to maintain the throughput. The research work focuses on selecting Keyframes, and the goal is to make a sequence of keyframes in the original sequence while reducing the redundancy as much as possible.

3.1. Partitioning of Video into Frames
Initially, image frames are extracted from a video. Generally, a video contains 24 image frames per second, out of which the majority are redundant frames. Therefore, there is a need to implement some measures to extract critical frames only. Usually, a video of k minutes is partitioned into:

Table 1. Text and Feature Extraction.

Sr.No.	Title of Research	Cite	Proposed Idea
1.	Movie Tags Prediction and Segmentation Using Deep Learning	[15]	Shot boundaries are computed by intersecting HSV (Hue, Saturation, Value) histograms of the successive frames.
2.	A Novel Key-Frames Selection Framework for Comprehensive Video Summarization	[16]	Key frames are identified by aggregating spatio temporal features generated by CapsuleNet.G1 continuity error and cubic H Bézier curve fitting for shot boundary detection within local sliding windows.
3.	A Novel Shot Detection approach using 8 Neighbors and Key-Colors	[17]	For eight neighbours, the grey difference is calculated between the adjacent pixels. If the threshold is exceeded, 1 will denote the corresponding statistic, and 0 otherwise. The main colours are applied to the adjacent frames using colour feathers to enhance the accuracy of shot detection and to judge the similarity of material between adjacent frames.
4.	An efficient method for video shot boundary detection and key frame extraction using SIFT point distribution histogram· Multi Modal Visual Features Based Video Shot Boundary Detection· Abrupt Shot Detection in Video using Weighted Edge Information· Shot boundary detection using perceptual and semantic Information	[18]	SIFT, SURF (Speed up Robust Features), Sobel Edge Detector, MSER (Maximally Stable Extreme Regions) are the key state-of-the-art local feature extraction algorithms used recently, using histograms-based feature extraction techniques.
5.	A Study of Discriminant Visual Descriptors for Sport Video Shot Boundary Detection	[19]	The most popular global attribute of a video are colour histograms. They offer a good balance between precision and time.
6.	Improving the Video Shot Boundary Detection Using the HSV Color Space and Image Subsampling	[20]	The colour space of the HSV is more random since it determines a 3-D colour that provides more robust detection results in turn.
7.	Edge Strength Extraction using Orthogonal Vectors for Shot Boundary Detection	[21]	Edge detection for feature extraction.
8.	Shot Boundary Detection in MPEG Videos Using Local and Global Indicators	[22]	Hybrid features for feature extraction
9.	Video shot transition detection using spatio temporal analysis and fuzzy classification	[23]	Image intensity or color histogram for feature extraction

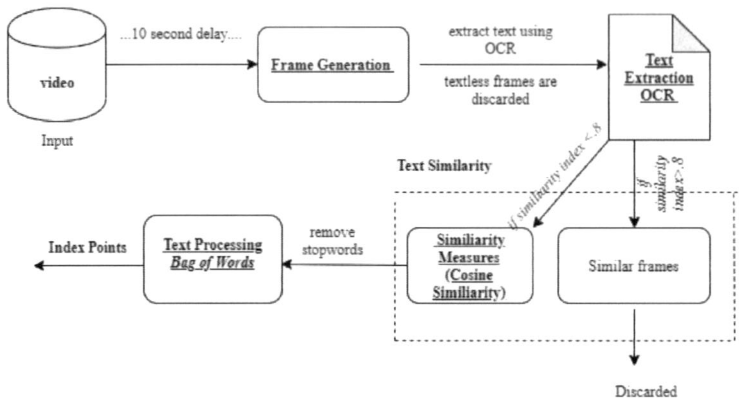

Figure 1. Index Point Detection Framework

$$N = k * 60 * 30 frames$$

where N is number of frames generated initially from the video.
It is evident that in educational videos, a topic will be discussed for at least 10 seconds. Therefore, to curtail the number of redundant frames, 10 seconds delay is introduced in frame generation. Figure 2 displays the frame generation of a sample video from the dataset after every 10 seconds.

Figure 2. Video Name: Data Structure. Total 85 frames are generated after introducing 10 Seconds delay

3.2. Extraction of Text using Optical Character Recognition

[24] Optical Character Recognition (OCR) tools use the binarization algorithm to identify text in image frames. At the same time, frames are containing plain text that often gives a

readable output. Frames with complex images, plots, or screenshots sometimes give incorrect and distorted output. To address this problem, we consider only those videos that have maximum content as plain text. Keyframes are detected after processing the text using the vectorization method. PyTesseract tool is used to extract text from image frames.

3.3. Cosine Similarity

Text similarity estimates the similarity index between the text documents using the distance measures or semantic similarity index. In this technique, text in the document acts as the local features used to find similarity between the documents. In our research, we will use the cosine similarity method [25].

Cosine similarity uses the cosine value of two vectors to represent the similarity between two image frames [26]. If frames are identical their cosine value is closer to 1, else varies from [0,1] as per equation 1 where A & B are image frames.

$$\cos(\mathbf{A},\mathbf{B}) = \frac{\mathbf{AB}}{\|\mathbf{A}\|\|\mathbf{B}\|} = \frac{\sum_{i=1}^{n}\mathbf{A}_i\mathbf{B}_i}{\sqrt{\sum_{i=1}^{n}(\mathbf{A}_i)^2}\sqrt{\sum_{i=1}^{n}(\mathbf{B}_i)^2}} \tag{1}$$

Values range from -1 and 1 , where -1 is perfectly dissimilar and 1 is perfectly similar. The similarity threshold is typically set between 0.7 and 0.8. It is observed that higher the compression ratio requirement, the smaller the threshold setting.

3.4. Index Point Identification

After successful extraction of image text, index points/keywords needs to be identified. The first step in this direction is text pre-processing [27], which involves removal of noisy data like punctuation marks, special character, tags, URL and removing redundant text components. In the second step removal of stop words viz. pronoun, prepositions, conjunctions and connecting verbs is done.

Third and final step are Text Preparation [28] in which the Bag of Words model is used to count word frequencies. The word frequency determines the possibility of a word to be called as *'Index Point'*.

In the process of Keywords extraction, removal of stop words, and conversion of the text into a list of words are also done. Unigram and bigrams are also generated so that accurate index points could be identified. These are converted to a matrix of integers by the process of vectorization. The figure 3 depicts that "Data structure", "Data type", "abstract data", and "primitive data", are bi-grams that are more often used by the instructor in the video.

4. Result Analysis

Keyframes extraction done using cosine similarity method with experimenting with different threshold values is shown in the table 2. The representation is only dramatic; actual comparisons are made for all videos dataset, as shown in table 3. The table reflects that the actual keyframes and the keyframes generated by our framework are almost equal, which justify that we achieve good precision at the threshold value of 0.8. Table 3 shows the precision and accuracy level for different videos as per the framework that we have developed. A suite of educational videos handpicked (source: NPTEL) is created. The keyframes with the high precision are identified which are indispensable for video summarization based on keywords.

It can be observed in figure 4 that almost all videos have a precision value of more than 70%. In some videos, due to the presence of images and mathematical formulas, the precision value is degraded. Those videos are selected willingly in the dataset to keep a check on the worst-case

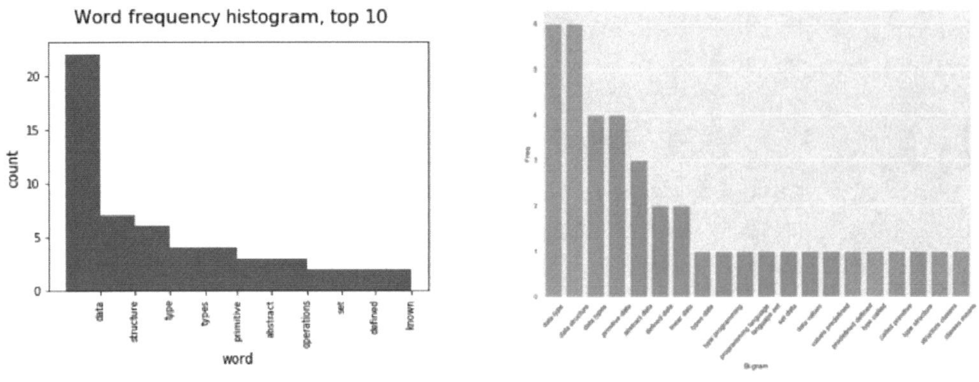

Figure 3. Histogram of most frequently occurring Uni-grams & Bi-Grams

Table 2. Threshold Comparison.

	Threshold 1	Threshold 2	Threshold 3	Threshold 4	Actual Key Frames
Threshold	>.5	>.6	>.7	>.8	0.8
Key frames (Data Structure video)	7	9	9	9	9
Key frames (PERT & CAM video)	18	19	20	24	22
Key frames (—-)	–	–	–	–	–

accuracy of the framework. The average precision value of the framework is 86%.

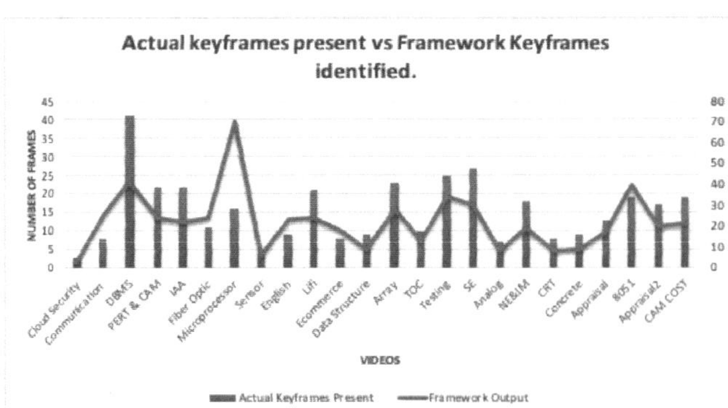

Figure 4. Actual Keyframes vs Algorithm Keyframes

Table 3. Video wise Precision table.

Video	Length	Total Frames	Videos satisfying Threshold	Original Keyframes (True Positive)	Precision Value
Cloud Security	9min38sec	58	3	3	1.00
DBMS	6min42sec	41	41	41	1.00
PERT & CAM	7min52sec	48	24	22	0.92
IAA	5min56sec	36	22	22	1.00
Sensor	9min39sec	61	6	5	0.83
Lifi	12min31sec	80	24	21	0.88
Ecommerce	6min32sec	40	18	8	0.44
Data Structure	14min4sec	85	9	9	1.00
Array	17min14sec	104	28	23	0.82
TOC	9min37sec	58	13	10	0.77
Testing	14min54sec	93	34	25	0.74
SE	12min23sec	79	30	27	0.90
Analog	9min59sec	60	8	7	0.88
NE&IM	7min47sec	47	19	18	0.95
CRT	9min58sec	60	8	8	1.00
Concrete	24min27 sec	154	9	9	1.00
Appraisal	15min54 sec	96	17	13	0.76
8051	17min55sec	109	40	19	0.48
Appraisal2	18min29 sec	111	20	17	0.85
CAM COST	11min53 sec	72	21	19	0.90

5. Conclusion

Through this research paper, we want to propose a new framework that extracts text from keyframes for keyword-based video summarization. Throughout the article, the keywords are referred to as Index Points. Cosine-Similarity approach is used to verify the index of similarity between the frames. OCR tool PyTesseract is used to extract text from the keyframes. The findings are rather persuasive, and the precision is considerably higher. This system would allow better indexing of videos in the future.

References

[1] Lu W, Sun H, Chu J, Huang X and Yu J 2018 *IEEE Access* **6** 40198–40211
[2] Mehmood I, Sajjad M, Rho S and Baik S W 2016 *Neurocomputing* **174** 393–403
[3] Ahanger G and Little T D 1996 *Journal of visual Communication and image representation* **7** 28–43
[4] Abowd G D 1999 *IBM systems journal* **38** 508–530

[5] Mahrishi M, Hiran K K, Meena G and Sharma P 2020 Machine learning and deep learning in real-time applications URL `https://www.igi-global.com/book/machine-learning-deep-learning-real/240152`
[6] Gharbi H, Bahroun S, Massaoudi M and Zagrouba E 2017 *2017 IEEE International Conference on Acoustics, Speech and Signal Processing (ICASSP)* (IEEE) pp 1502–1506
[7] Asim M, Almaadeed N, Al-Máadeed S, Bouridane A and Beghdadi A 2018 *2018 Colour and Visual Computing Symposium (CVCS)* (IEEE) pp 1–6
[8] Tarare P and Jadhav D 2015 *2015 International Conference on Applied and Theoretical Computing and Communication Technology (iCATccT)* (IEEE) pp 403–408
[9] Ejaz N, Tariq T B and Baik S W 2012 *Journal of Visual Communication and Image Representation* **23** 1031–1040
[10] Duan Y, Lu J, Feng J and Zhou J 2018 *IEEE transactions on pattern analysis and machine intelligence* **40** 1139–1153
[11] Mahrishi M and Morwal S 2020 *Advances in Intelligent Systems and Computing AISC Springer*
[12] Mademlis I, Tefas A and Pitas I 2018 *Information Sciences* **432** 319–331
[13] Liu H, Lu H and Xue X 2012 *IEEE transactions on knowledge and data engineering* **25** 1706–1718
[14] Lanying H and Cong-huan Y 2012 *2012 2nd International Conference on Consumer Electronics, Communications and Networks (CECNet)* (IEEE) pp 2179–2182
[15] Khan U A, Martínez-Del-Amor M Á, Altowaijri S M, Ahmed A, Rahman A U, Sama N U, Haseeb K and Islam N 2020 *IEEE Access* **8** 6071–6086
[16] Huang C and Wang H 2019 *IEEE Transactions on Circuits and Systems for Video Technology*
[17] Liu H B, Tan T H, Huang S C, Wang Z J and Zhou J 2018 *2018 7th International Symposium on Next Generation Electronics (ISNE)* (IEEE) pp 1–3
[18] Tippaya S, Sitjongsataporn S, Tan T, Khan M M and Chamnongthai K 2017 *IEEE Access* **5** 12563–12575
[19] Tippaya S, Tan T, Khan M and Chamnongthai K 2015 *2015 10th Asian Control Conference (ASCC)* (IEEE) pp 1–4
[20] Liu F and Wan Y 2015 *2015 Seventh International Conference on Advanced Computational Intelligence (ICACI)* (IEEE) pp 351–354
[21] Priya G L and Domnic S 2012 *Procedia Technology* **6** 247–254
[22] Ren J, Jiang J and Chen J 2009 *IEEE Transactions on Circuits and Systems for Video Technology* **19** 1234–1238
[23] Lin G S, Chang M K and Chiu S T 2009 *International Journal of Pattern Recognition and Artificial Intelligence* **23** 1179–1200
[24] Kaundilya C, Chawla D and Chopra Y 2019 *2019 6th International Conference on Computing for Sustainable Global Development (INDIACom)* (IEEE) pp 145–150
[25] Hoque M N, Islam R and Karim M S 2019 *2019 1st International Conference on Advances in Science, Engineering and Robotics Technology (ICASERT)* (IEEE) pp 1–5
[26] Ma M, Met S, Hou J, Wan S and Wang Z 2017 *2017 International Symposium on Intelligent Signal Processing and Communication Systems (ISPACS)* (IEEE) pp 164–169
[27] Yao J 2019 *Journal of Physics: Conference Series* vol 1187 (IOP Publishing) p 052020
[28] Sarwar A, Mehmood Z, Saba T, Qazi K A, Adnan A and Jamal H 2019 *Journal of Information Science* **45** 117–135

Analytical Study of Blockchain Enabled Security Enhancement Methods for Healthcare Data

Moumita Roy[1,2], Dr. Monisha Singh[3]

[1]PhD Scholar, Department of Computer Science, Christ Deemed to be University, Bengaluru, India

[2]Assistant Professor, Department of Computer Application, CMR Institute of Technology, Bengaluru, India

[3]Assistant Professor, Department of Computer Science, Christ Deemed to be University, Bengaluru, India

moumita.r@cmrit.ac.in

Abstract. Healthcare data protection is very much required in the present time due to the rate of growth in data volume and sensitivity of the data. The data is collected through EHR, diagnostic image like scan or x-ray or IoT wearable devices which has sensors and shared with stakeholders who can analyse the data and use it for betterment of the processes. Benefits of medical data sharing among the entities makes it even more vulnerable. In this paper we have studied and analysed few blockchain based methods of securing healthcare data which is stored in cloud. The methods studied is successful in keeping the data safe from unauthorized access.

Keywords: Blockchain, Healthcare, Cloud, Ethereum, Hyperledger Fabric, EHR, IoT

1. Introduction

The healthcare industry has undergone a large-scale rapid-paced digitization transformation. This positive technological impact generated a massive load of electronic records like the Electronic Medical Report (EMR) or Electronic Healthcare Data (EHR) that needs to be managed and accessed by the stakeholders. Availability and security of these data sources is extremely important as it directly affects the health and lives of the patients and their treatment. Also, such data can be used for research – related work or analyzed to develop market strategies.

Cloud storage is a convenient solution for storing huge bulk of data in the present times. The ubiquitous, economic, scalable and on-demand features of cloud storage make it more efficient as a user data repository as compared to other traditional storage mechanisms. The concept of cloud computing can be further extended to decentralized idea of fog computing. Cloud/fog computing can work with other technologies, like IoT networks, at ease [1]. A major limitation of cloud storage is that the users cannot control the data they stored on cloud and they have to rely on the service providers for keeping the data secure. External dependency makes the data extremely vulnerable and the issues of user trust, data accessibility and security need to be solved [2].

Decentralized cloud solutions can be an answer to this data vulnerability issue. Smart contracts [3] along with blockchain [4] can create traceability and provide secure data access in cloud platforms [5].

Content from this work may be used under the terms of the Creative Commons Attribution 3.0 licence. Any further distribution of this work must maintain attribution to the author(s) and the title of the work, journal citation and DOI.
Published under licence by IOP Publishing Ltd

A contract can be created, verified and applied between users, using rules of smart contract, to ensure the execution of a fair data transaction. Blockchain decentralizes the network, makes it more stable and stops any unauthorized mediatory intervention. That way, the data on the cloud platform becomes more secure. The merger of blockchain and cloud technologies have shown a lot of potential in the recent times in managing data access. The reason behind the success of the blockchain technology in enhancing security is its complexity [6]. A lot of effort is needed to understand the complex technology, create and implement a blockchain, and that makes it almost impossible to break it.

With the increase in the number of e-health services, volume of health-related documents being stored on cloud is also on a rise. Health records are confidential and requires secure storage. To mitigate the security challenges of cloud storage and provide uninterrupted and economic healthcare services, blockchain assisted cloud storage can be an excellent solution. We have identified few such research solutions which are quite promising and mature in their approach. Our study aims at analyzing and comparing them, bringing forward their strengths and weaknesses and also bringing to light the research gaps, if any.

2. Blockchain Technology

The blockchain concept was invented by Satoshi Nakamoto in the year 2008 for building cryptocurrency bitcoin [7]. Blockchain is a distributed public transaction ledger which keeps growing every time a new transaction takes place and a new record or block is added to the existing chain. The new block contains cryptographic hash of the previous block, time of its creation and other transactional data that are encoded into a Merklee tree. A record once added to a blockchain, cannot be modified.

The primary reasons that make a blockchain extremely secure are its de-centralized nature, immutability, data encryption and change transparency [8]. Traditional centralized systems like the client-server store all data in a centralized repository, making the system vulnerable to hacking. During maintenance or downtime, the entire system becomes unavailable and in worst case if the systems gets corrupt or fails beyond repair, all the data will be lost. Blockchain being a decentralized network, is almost not hackable. It has the benefit of not being controlled by any central authority or third-party controller. Blockchain is immutable means the data once stored in a blockchain becomes tamper proof. This can be achieved using cryptographic hash functions. Blockchain is transparent but it also protects the identity of the user. Transaction history will show each and every transaction performed but the username will be encrypted for protecting user's privacy.

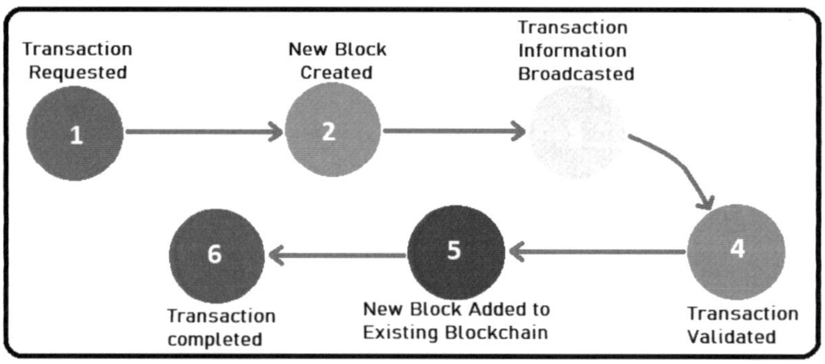

Figure 1. Blockchain Architecture work-flow for a new transaction.

When a new transaction is initiated by a user, a new block is created containing the encrypted data and link to previous block. The information about the transaction is passed on to all users and the transaction is validated. The validated transaction block then gets connected to the existing chain of blocks. Fig. 1 depicts the flow of creation of a new block in a blockchain [9].

The use of blockchain started with the cryptocurrency Bitcoin but now it is being used for making smart contracts, creating supply chain, network security enhancement, banking and many more industry sectors. Presently many blockchain platforms are widely in use, like Ethereum [10], Hyperledger Fabric [11] and IBM Blockchain [12].

Ethereum and Hyperledger Fabric are the two most commonly used public blockchain smart contract platforms. Ethereum technologies run on Ethernet Virtual Machine (EVM) and uses a public network. Each node in this network is identifiable by all other nodes. Ethereum users communicate with other nodes using smart contracts. The miner node solves the decryption puzzle (Proof of Work) and earns a reward (called Gas). Hyperledger Fabric is a permissioned blockchain in which nodes cannot publicly identify each other. Each node can be configured to perform specific tasks. An endorsing node sends transaction information to few or all other nodes. The endorsing node simulates the execution of the transaction and sends the result to the client node. The client node sends the result to the ordering node which creates the block for that transaction. For multiple transactions, node creation is done on first come first serve basis.

3. Healthcare and Technology

Healthcare now is of importance like never before, due to the increasing global population, the rise in the average age of the global population and the chronic and newer lifestyle disease that the world is getting affected with every day. Given the situation, more and more people will become old and suffer from chronic diseases and as a result basic healthcare will become super expensive and out of reach of common people. To get out of this deadlock, technological intervention is required.

Technological aid cannot reduce the rate of population aging or their getting sick with chronic illness, but can make the healthcare services affordable and accessible. Technology can provide with the following light-on-the-pocket healthcare advances which can save many lives:

- Monitoring a patient in the real-time and reporting any change in the status immediately.
- The end-to-end process of treating a patient can be automated making data flow from one equipment to the other with the help of wireless communication. That way the process will be much faster and error free.
- Data can be analyzed to generate reports directly, without storing the raw data. A lot of memory space will be saved.
- Devices can be configured to set alert to medical personnel for patients who need critical care or are on life support.
- Online medical consultations can be possible when the patient and the medical practitioner are geographically apart.
- Data collected can be used in research work and improvement of the future.

Technologies like Artificial Intelligence, Cloud Computing, Big Data and Internet of Things (IoT) are helping in automating healthcare processes by connecting the medical devices over a network and provide scalable virtual infrastructures for computation and storage of data and reports.

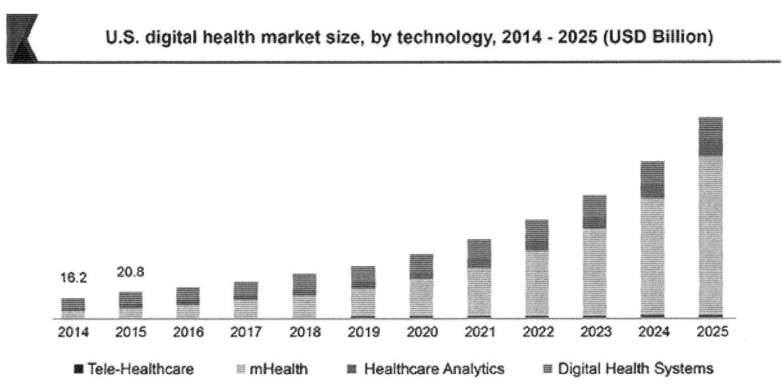

Figure 2. Growth in Digital Health Market in U.S. between 2014 to 2025. [9].

Across the globe, the use of technology in Healthcare domain is rapidly growing over the past decade. Fig. 2 shows the statistics of the rate of digitization in U.S.

4. Blockchain Based Healthcare Applications

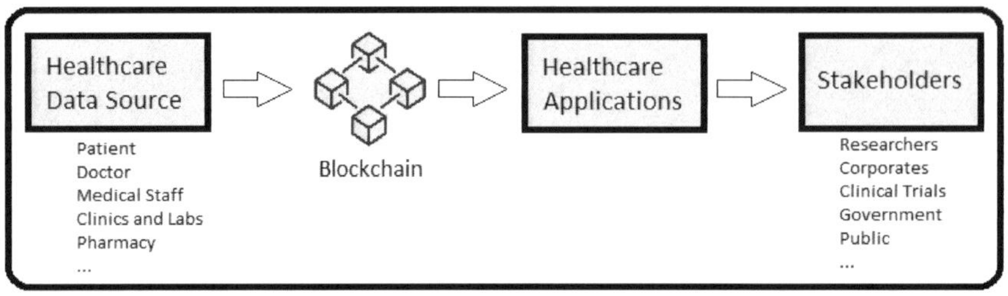

Figure 3. Blockchain implementation in Healthcare applications – the workflow.

After digitization of healthcare, a huge volume of medical data is getting generated. These data are critical and sensitive; critical as the data needs to be always available for the timely medical assistance of the patient and sensitive as the data can be very personal to the person involved. These data need to be protected and kept safe from any manipulation or loss.

Blockchain technology can be used to provide protected storage to the healthcare data generated due to digitization. Presently many healthcare applications have implemented blockchain in their system [13]. Such systems start with the generation of the data which after getting validated using a blockchain is send to the medical application through which the data reaches the concerned stakeholders. Raw data collected from primary sources like the patients, medical practitioners and staffs, pharmacies and pathological labs are gathered which becomes the starting point of the data management operation (example, Electronic Health Record). Blockchain technology secures this collected data using encryption and consensus algorithms. This secure data is accessed by the stakeholders like pharmaceutical companies, researchers, government authorities etc. through the healthcare applications. Fig. 3 shown the diagrammatic representation of the process flow.

In this paper, we will concentrate primarily on the data gathering, blockchain validation and storage of the healthcare data. The entire process of collecting and protecting the data, storing it and making is accessible is very much needed as the healthcare related data is critical and can be put to multiple and extremely important uses. Regarding security, as discussed, blockchain can provide a productive and robust solution.

.

4.1. Why sharing of medical data is important?

Traditional healthcare system requires a lot of human interference and judgement. If the system has to be automated with smart components, the system has to collect, store, learn from and manage a huge bulk of data. By doing so the practitioners can provide better care to the patients and collaboration among the stakeholders will improve the overall system. This sharing of information can happen in multiple levels and dimensions. A patient can share his medical history with doctor and insurance company, doctor can share data with pharmaceutical company and pathological laboratory, pharmaceutical company can interact with research centre and so on [14].

The types and formats of data shared are varied. The data can be in form of EHRs, medical images, data coming from personal wearable devices or XML files (CDA). These data can also be either static data or dynamic data. Static data are those that do not change much like patient's name, address, height (for adults) etc. Dynamic data is more critical data like blood sugar level, level of oxygen in blood, blood pressure etc. [15]

Along with advantages, there are certain limitations of sharing data. Patients do not have access or control over the EHR after sharing and they do not get to know which unknown entities are using their data. As medical data contains lot of personal information, keeping it secure is very important.

4.2. Storage of Medical Data

As medical data cumulatively forms a huge bulk, cloud storage is mostly preferred due to its convenience. Cloud storage provides with a cost effective and virtually endless storage space with hindrance-free accessibility and fast transmission of data. With the invent of Mobile Cloud Computing (MCC) and its ability to fetch data from wearable medical devices, data can be analyzed very fast and timely reporting is possible [16]. The continuous flow of dynamic data can be accommodated in the storage system using several cloud environments like Amazon Web Services (AWS) or Google Cloud The limitation with cloud storage is its data securing capabilities as dependency lies with third party service providers and unauthorized access of data cannot be avoided totally.

4.3 Securing Medical Data using Blockchain

Insecure medical data can be a loss of personal data to the patient and also can cause harm to a medical institute's reputation and wealth. Blockchain can provide data security with its decentralized structure, applying cryptography and making data immutable. Consensus mechanism of blockchain can authorize the data requesting entity, securing the data from unauthorized access [14]. Also, the encrypted data will be meaningless to any intermediate entity except the data owner. Hacking and decrypting the data will be extremely difficult (almost impossible) owing to the complexity of blockchain technology [17]. The decentralized structure and use of smart contracts can also track the behaviour of the data elements and in case of any protocol violation, user access rights can be revoked [18].

5. Analysis

We have compared number of methods for securing healthcare data stored in cloud environment using blockchain technology. Table 1 contains a detailed comparison of those methods. Few methods used EHR as the input data source whereas few uses diagnostic images or reads input data from wearable medical devices or mobile devices. Majority of the methods are using Ethereum as the blockchain platform till date, but Hyperledger Fabric is slowly gaining popularity. Most of the methods using blockchain is also using cloud environment to store the data. Cloud platforms like AWS, Google cloud are preferred.

The methods we studied in this paper are all able to enhance data security by controlling access and stop unauthorized access using blockchain technology. Xia et al. [15] claimed that the proposed algorithm can keep track of the behaviour of the data and in case there is any protocol violation, the algorithm can revoke access rights for that data unit. This advancement can help keeping data secure as a continuous process. Kaur et al. [18] in their work could store both static and dynamic medical data in

Table 1. Comparing Multiple Healthcare Data Management Methods Using Blockchain

the blockchain, which is also a commendable advancement. But during data transmission, data loss due to leakage is a concern in this process.

Article	Input Data Format	Blockchain Platform	Other Technologies	Benefits	Drawbacks	Gap Identified
13	HER	Ethereum	AWS, IPFS	Faster sharing of medical data over mobile cloud Can avoid	Not very scalable Poor network latency Compromised data privacy	Undefined

				unauthorized data access	while uploading	
14	HER	Ethereum	Undefined	Access control and authentication Secure search Privacy preservation Collision resistance	Higher data access communication cost The longer the data package, the more is the time cost to send transaction to blockchain	Implement proposed algorithm using Hyperledger Fabric
15	Cloud Stored Medical Data	Undefined	JMeter	Data access control Revoke data access on rule violation	Network Latency increases in proportion with number of requests from cloud service provider	Undefined
16	HER	Undefined	KSI, Jmeter	Better time and cost compared to other existing solutions	Undefined	Implement solution on smaller data stored on cloud
17	Patient diagnostic data	Ethereum	ECC	Block ID generation cost is same for all input size Successful privacy protection	The longer the data package, the more is the time cost to send transaction to blockchain Encryption takes way longer than decryption	Interoperability among other medical stakeholders Include techniques for key loss
18	HER	Undefined	Cloud Environment	Securely stores heterogeneous data	Undefined	Blockchain based solutions for enterprise data warehouse issues
19	Static and	Ethereum	AWS, Google Cloud	Divides input data into static	After the data transfer, any	After the data

	dynamic personal health data			and dynamic Data quality is maintained using machine learning	data leakage while the data is stored or in use is out of scope.	transfer, any data leakage while the data is stored or in use can be included in scope.
20	Medical Image	Custom-designed blocks	Undefined	Stronger data protection. Time taken does not depend on the number of keywords	Efficiency is slightly lower	Improved efficiency
21	Personal health data from wearable devices	Ethereum	Undefined	Improved efficiency	Transactions generated using simulators	Algorithms to train the system for independent decision making in the real time Real transactions between devices
22	CDA	Hyperledger Fabric	XDS, FHIR	Secure immutable records Scalable	Undefined	Undefined

6. Conclusion

Medical data volume is on a rise due to increase in the medical IoT devices and automation in the healthcare processes. Mostly these data are stored in form of EHR, EMR or diagnostic images. These data are extremely critical for timely treatment of the patient and also it contains personal and sensitive information of the patient. Due to the bulk, the data is mostly stored in cloud environments. Cloud is easy to access, virtually endless, always accessible, scalable and pocket-friendly; so, it has become the most convenient option for storing data. But cloud has security issues. To overcome this issue and keep the healthcare data safe from intruders, blockchain technology is used. Here we have studied few of the proposed solutions of recent times. The data security requirements are met, but the solutions have few limitations and gaps which require attention in the future.

References:

[1] M. Yannuzzi, R. Milito, R. Serral-Gracià, D. Montero and M. Nemirovsky, "Key ingredients in an IoT recipe: Fog Computing, Cloud computing, and more Fog Computing," 2014 IEEE 19th International Workshop on Computer Aided Modeling and Design of Communication Links and Networks (CAMAD), Athens, 2014, pp. 325-329, doi: 10.1109/CAMAD.2014.7033259.

[2] A. Awad, A. Matthews, Y. Qiao and B. Lee, "Chaotic Searchable Encryption for Mobile Cloud Storage," in IEEE Transactions on Cloud Computing, vol. 6, no. 2, pp. 440-452, 1 April-June 2018

[3] N. Szabo, "Smart Contracts: Building Blocks for Digital Markets," EXTROPY: The Journal of Transhumanist Thought, 1996.

[4] S. Nakamoto, "Bitcoin: A Peer-to-Peer Electronic Cash System"; http://bitcoin.org/bitcoin.pdf, 2008.

[5] D. K. Tosh, S. Shetty, X. Liang, C. A. Kamhoua, K. A. Kwiat and L. Njilla, "Security Implications of Blockchain Cloud with Analysis of Block Withholding Attack," 2017 17th IEEE/ACM International Symposium on Cluster, Cloud and Grid Computing (CCGRID), Madrid, 2017, pp. 458-467, doi: 10.1109/CCGRID.2017.111.

[6] W. Zheng, Z. Zheng, X. Chen, K. Dai, P. Li and R. Chen, "NutBaaS: A Blockchain-as-a-Service Platform," in IEEE Access, vol. 7, pp. 134422-134433, 2019, doi: 10.1109/ACCESS.2019.2941905.

[7] Nakamoto, Satoshi. (2009). Bitcoin: A Peer-to-Peer Electronic Cash System. Cryptography Mailing list at https://metzdowd.com.

[8] https://blockgeeks.com/guides/what-is-blockchain-technology/

[9] Mahrishi, M., Hiran, K. K., Meena, G., & Sharma, P. (2020). Machine Learning and Deep Learning in Real-Time Applications. IGI Global.

[10] Wood, Gavin. "Ethereum: A secure decentralised generalised transaction ledger." Ethereum project yellow paper 151.2014 (2014): 1-32.

[11] Elli Androulaki, Artem Barger, Vita Bortnikov, Christian Cachin, Konstantinos Christidis, Angelo De Caro, David Enyeart, Christopher Ferris, Gennady Laventman, Yacov Manevich, Srinivasan Muralidharan, Chet Murthy, Binh Nguyen, Manish Sethi, Gari Singh, Keith Smith, Alessandro Sorniotti, Chrysoula Stathakopoulou, Marko Vukolić, Sharon Weed Cocco, and Jason Yellick. 2018. Hyperledger fabric: a distributed operating system for permissioned blockchains. In Proceedings of the Thirteenth EuroSys Conference (EuroSys '18). Association for Computing Machinery, New York, NY, USA, Article 30, 1–15. DOI:https://doi.org/10.1145/3190508.3190538

[12] Smith, Ben, and Konstantinos Christidis. "IBM Blockchain: An enterprise deployment of a distributed consensus-based transaction log." Proc. Fourth International IBM Cloud Academy Conference. 2016.

[13] D. C. Nguyen, P. N. Pathirana, M. Ding and A. Seneviratne, "Blockchain for Secure EHRs Sharing of Mobile Cloud Based E-Health Systems," in IEEE Access, vol. 7, pp. 66792-66806, 2019, doi: 10.1109/ACCESS.2019.2917555.

[14] Y. Wang, A. Zhang, P. Zhang and H. Wang, "Cloud-Assisted EHR Sharing With Security and Privacy Preservation via Consortium Blockchain," in IEEE Access, vol. 7, pp. 136704-136719, 2019, doi: 10.1109/ACCESS.2019.2943153.

[15] Q. Xia, E. B. Sifah, K. O. Asamoah, J. Gao, X. Du and M. Guizani, "MeDShare: Trust-Less Medical Data Sharing Among Cloud Service Providers via Blockchain," in IEEE Access, vol. 5, pp. 14757-14767, 2017, doi: 10.1109/ACCESS.2017.2730843.

[16] Nagasubramanian, Gayathri & Sakthivel, Rakesh & Patan, Rizwan & Gandomi, Amir & Muthuramalingam, Sankayya & Balamurugan, Balamurugan. (2018). Securing e-health records using keyless signature infrastructure blockchain technology in the cloud. Neural Computing and Applications. 10.1007/s00521-018-3915-1.

[17] Omar, Abdullah & Bhuiyan, Md & Basu, Anirban & Kiyomoto, Shinsaku & Rahman, Shahriar. (2019). Privacy-friendly platform for healthcare data in cloud based on blockchain environment. Future Generation Computer Systems. 95C. 511-521. 10.1016/j.future.2018.12.044.

[18] Kaur, Harleen & Alam, Afshar & Jameel, Roshan & Mourya, Ashish & Chang, Victor. (2018). A Proposed Solution and Future Direction for Blockchain-Based Heterogeneous Medicare Data in Cloud Environment. Journal of Medical Systems. 42. 156. 10.1007/s10916-018-1007-5.

[19] X. Zheng, R. R. Mukkamala, R. Vatrapu and J. Ordieres-Mere, "Blockchain-based Personal Health Data Sharing System Using Cloud Storage," 2018 IEEE 20th International Conference on e-Health Networking, Applications and Services (Healthcom), Ostrava, 2018, pp. 1-6, doi: 10.1109/HealthCom.2018.8531125.

[20] Zhang, Yu-lei & Wen, Long & Zhang, Yong-jie & Wang, Cai-fen. (2020). Deniably authenticated searchable encryption scheme based on Blockchain for medical image data sharing. Multimedia Tools and Applications. 10.1007/s11042-020-09213-w.

[21] Akkaoui, Raifa & Hei, Xiaojun & Cheng, Wenqing. (2020). EdgeMediChain: A Hybrid Edge Blockchain-Based Framework for Health Data Exchange. IEEE Access. PP. 1-1. 10.1109/ACCESS.2020.3003575.

[22] Margheri, Andrea & Masi, Massimiliano & Miladi, Abdallah & Sassone, V. & Rosenzweig, Jason. (2020). Decentralised Provenance for Healthcare Data. International Journal of Medical Informatics. 141. 104197. 10.1016/j.ijmedinf.2020.104197.

A Review on Segmentation of Vitiligo image

Neha Agrawal[1,3], Dr. Sagaya Aurelia[2]

[1]PhD Scholar, Department of Computer Science, Christ Deemed to be University, Bengaluru, India & Assistant Professor, Department of Computer Application, CMR Institute of Technology, Bengaluru, India

[2]Assistant Professor, Department of Computer Science, Christ Deemed to be University, Bengaluru, India

neha.ag@cmrit.ac.in

Abstract. Digital image processing is playing a vital role in medical science. Segmentation is a very important and critical step of image processing to analyse the image. Segmentation has become an eminent objective in image analysis and computer vision algorithms. Vitiligo is a very common skin disease that causes depression for the patient suffering from it. In current work, several segmentation techniques like thresholding, watershed, Region-based RAG, Felzenszwalb Segmentation, SLIC - K-Means based image segmentation, and Quickshift image segmentation were analysed along with their experimental results on vitiligo image. The execution time of different algorithms was also measured and analysed. Some of the advantages and disadvantages of applied techniques were discussed.

Keywords: Vitiligo, Segmentation, Image Processing, Thresholding, Watershed

1. Introduction

Skin disease is a very common problem faced by many people in the world. There is a fatal and non-fatal skin disease. Vitiligo is a very common skin disease and does not lead to death. In this disease white patches appear on the skin. These patches are caused by lack of a cell melanin which is responsible for skin color. Melanin is produced by melanocytes cell. It is an autoimmune disease where melanocytes cell is mistakenly killed by the immune system. However, it is not harmful but it affects the mental state of humans. They lose confidence to go in public. Vitiligo treatment is time-consuming. It takes several visits for a patient to get it diagnosed, and also accurate identification of disease depends on the year of experience of the doctor.

The influence of digital image processing in the modern world is enormous. Digital image processing plays an important role in the diagnosis of various diseases in the medical field. Digital image processing consists of different steps to process the image. These steps include image procurement, pre-processing, segmentation, and classification. Every step in image processing is important, but image segmentation is a very critical step. The good quality segmented image leads to better classification accuracy. Digital image processing steps are shown in Figure1.

Content from this work may be used under the terms of the Creative Commons Attribution 3.0 licence. Any further distribution of this work must maintain attribution to the author(s) and the title of the work, journal citation and DOI.
Published under licence by IOP Publishing Ltd

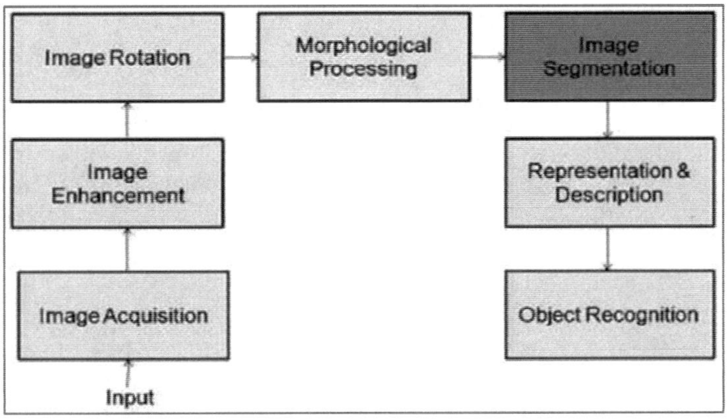

Figure 1: Steps of Digital Image Processing Source: Adapted from [11]

Image segmentation is the process of partitioning the image into various segments to analyze the different parts of the image. Each segment shares the same characteristics. Later interested segments can be extracted from an image for further analysis. There are several techniques available for image segmentation. In this work, we will discuss different segmentation techniques which can be applied to vitiligo image.

2. Related work

Liu used CNN in HSV and YCrCb for lesion detection which was an effective mechanism [1]. The author used MFCC and i-vectors to extract characteristics of images [2].]. ASLM techniques consist of 4 steps for the segmentation of melanoma images [3]. In another work Gradient and feature, the adaptive contour was used for identification of melanoma image [4]. Segmentation was inbuilt in the LVQ neural network for vitiligo images [5]. Various segmentation techniques review was presented by Gurusamy[6]. In another work, CNN with U Net architecture and watershed algorithm was used for automatic segmentation of vitiligo images [7]. The graphical interface was developed and face contouring, patched ROIs definition, image filtering selection, and global threshold setting were used [8]. Local binary pattern, GLCM, DCT, and DFT is one of the techniques for extracting features of images [9]. In another work, WLD was used along with GLCM, DCT, DFT, and LBP for feature extraction of vitiligo images. Density-based and active contour approaches were used to detect the border of skin lesions [12]. GLCM was used for feature extraction [13]. In other work image inpainting, canny based edge detection, the morphological operation was used for image segmentation [14].

3. Segmentation techniques

3.1: Thresholding: Thresholding can be used as an image segmentation technique, where pixels of the image can be changed to easily analyze the image. In thresholding, a grayscale image is converted into a binary image.
(i) Normal Thresholding:
Otsu's segmentation: Global Otsu's segmentation works on the bimodal histogram of the image. The bimodal histogram is the histogram of the frequency of the pixel intensity of the image. Here threshold will be calculated automatically from the histogram. This works on the concept of minimizing the intraclass intensity variance and maximizing the inter-class intensity variance.
If the background of the image is relatively uniform, then the global threshold gives good results. However, if there is more change in the background intensity, adaptive thresholding (a.k.a. local or dynamic thresholding) may give better results. However local is much slower than global thresholding.

(ii) Multi-Otsu Thresholding. The multi-Otsu threshold is a thresholding algorithm that is used to separate the pixels of an input image into several different classes, each one obtained according to the intensity of the gray levels within the image.

*3.2: Watershed segmentation.*The watershed algorithm works on grayscale image boundary between two regions are calculated based on the brightness of pixels. The image is viewed as a landscape where bright pixels formed a high peak. With the help of markers, this landscape is flooded until different flood basins meet at peaks. Each different basin is considered as different regions. Compactness feature can be added for the region with regular shape.

*3.3 Region-based RAG.*This algorithm works on a region adjacency graph which represents the adjacency of regions with a graph. Each region is considered as a node in the graph. The edge between the two regions will be given weight which can be defined based on the difference in average color between the two regions.

*3.4 Felzenszwalb Segmentation.*It uses a minimum spanning tree-based clustering for image segmentation of multichannel image (RGB image). The image can be smoothened before segmentation with the help of the Gaussian kernel. Several segments can be decided by the user. The algorithm uses Euclidean distance between pixels in color space.

*3.5 SLIC - K-Means based image segmentation.*It is similar to the quickshift algorithm where it uses the K-means algorithm in the 5d space of color information and image location. It is very simple and efficient. Several segments can be provided by the user.

*3.6 Quickshift image segmentation.*This algorithm is a 2D image segmentation algorithm and represents the family of the local mode-seeking algorithm. It simultaneously computes hierarchical segmentation on multiple scales. This scale shows the local density approximation. Also, the level can be selected to produce hierarchical segmentation.

4. Experimental Results

Image for vitiligo was collected from kaggle.com which is the online repository of the dataset used for the machine learning algorithm. We have used python with Jupiter notebook to implement the different algorithms. Color image of vitiligo as shown in Figure 2 was converted into a grayscale image. Then normal thresholding was applied to the image.

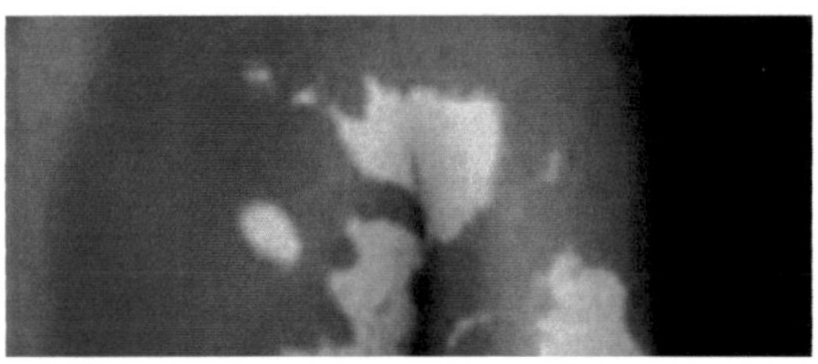

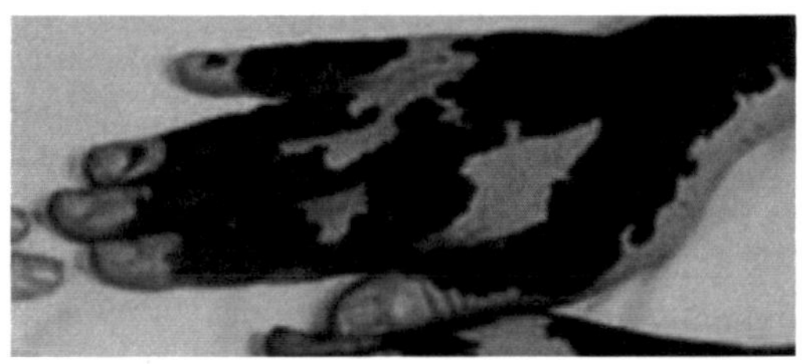

Figure 2. Color Images of Vitiligo.

First, we have applied regular thresholding and inverted thresholding to the image where pixels were converted to either black or white color depends on the threshold value. This thresholding works well with the binary image. We have set the value of the threshold to 127. The result is shown in figure 3. As can be seen from the result that this algorithm did not work well with our data set.

Figure 3: Result of Manual Threshold and Inverted Threshold.

The next threshold method used was global otsu's thresholding which is based on the bimodal histogram. Here the value of the threshold was calculated automatically based on the histogram. Then the threshold was applied to the image. This calculated threshold can be seen as a red line in the histogram of the figure. As it can be seen from figure 4 that otsu's thresholding gives a better result than manual thresholding, however, the result is not promising still.

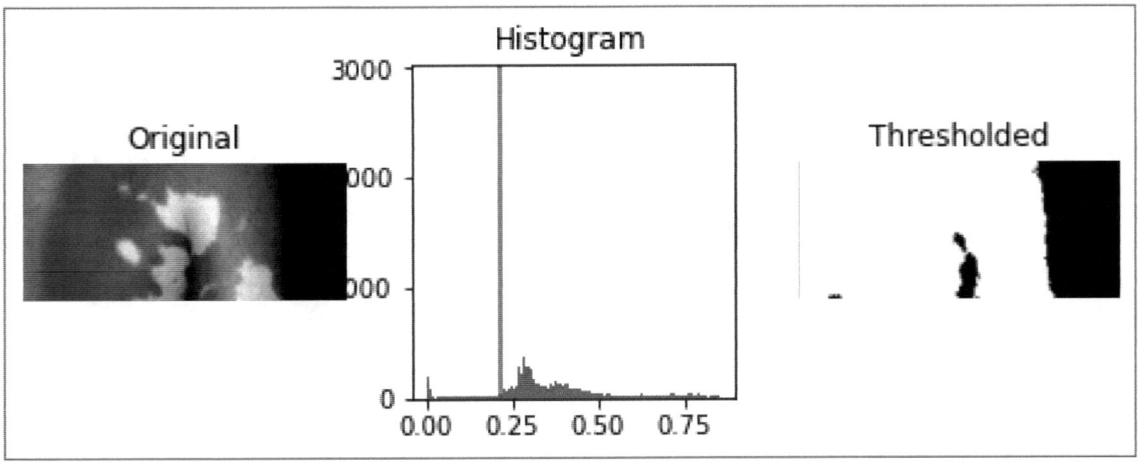

Figure 4: Result of Global Threshold.

Next local thresholding was applied on the vitiligo image. For each pixel, an optimal threshold was defined by maximizing the variance between two classes of pixels of the local neighborhood. This color image of vitiligo was converted into a grayscale image. Then the image was converted into an 8-bit unsigned integer format. The value of the radius was taken as 15. Figure 5 shows the comparison of local threshold and global threshold on the vitiligo image.

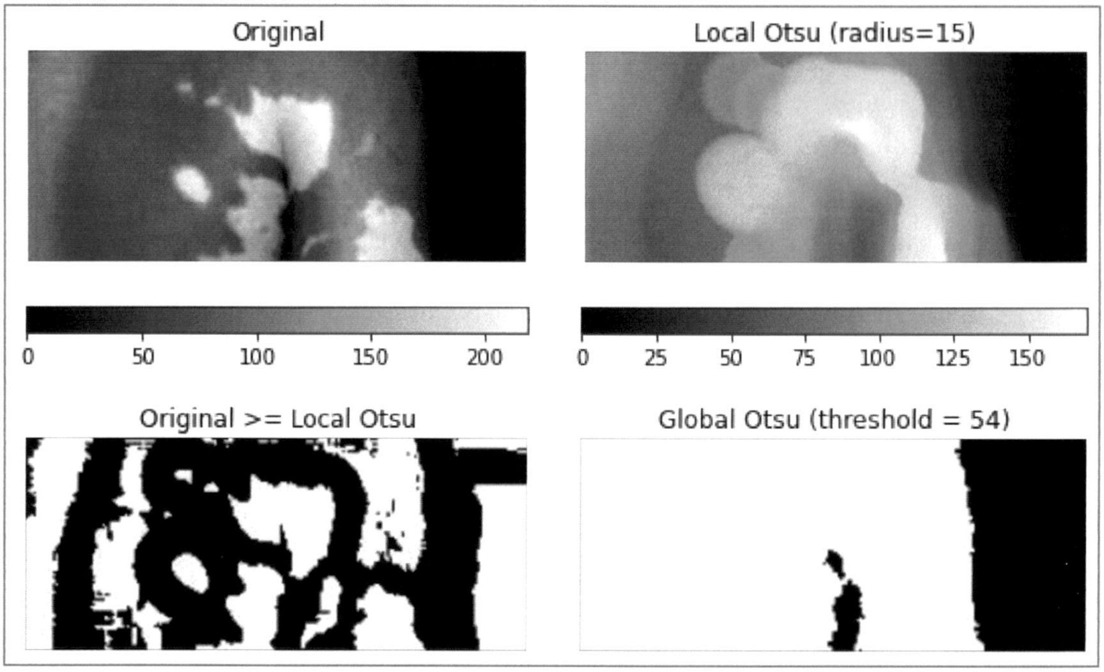

Figure 5: Comparison of Local Threshold & Global Threshold.

Multi-Otsu calculates several thresholds, determined by the number of desired classes. The number of classes taken here is 3: for obtaining three classes, the algorithm returned two threshold values. They are denoted by a red line in the histogram given below in Figure 6.

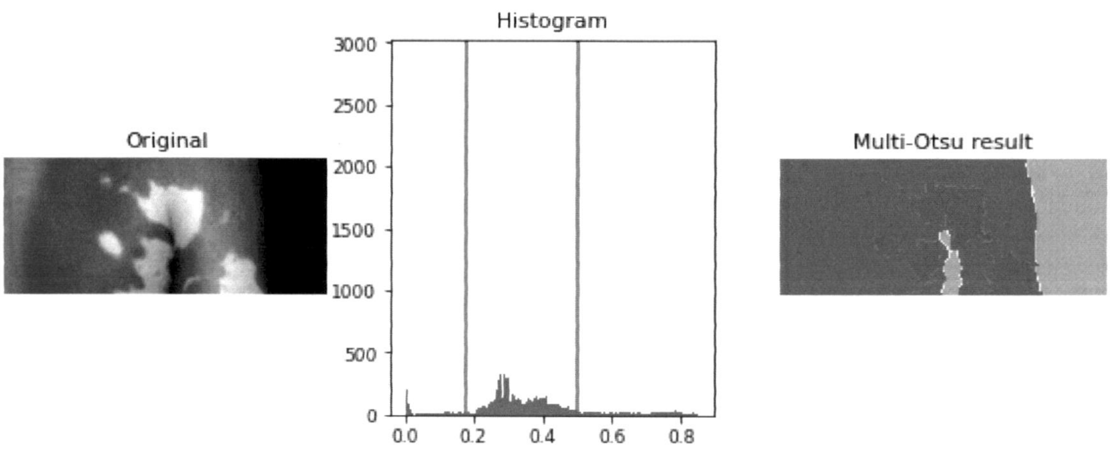

Figure 6: Result of Multi-Otsu.

The next watershed algorithm was used to segment the vitiligo image. The first color image was converted into a grayscale image. The image was de-noised and the local gradient was calculated with the value of radius taken as 5 to get the smoother image. Then a marker image was constructed from the region of the low gradient within the image with the condition gradient less than 10. In a gradient image, the areas of high values provide a blockade that helps to segment the image. Using markers on the lower values will ensure that the segmented objects are found.

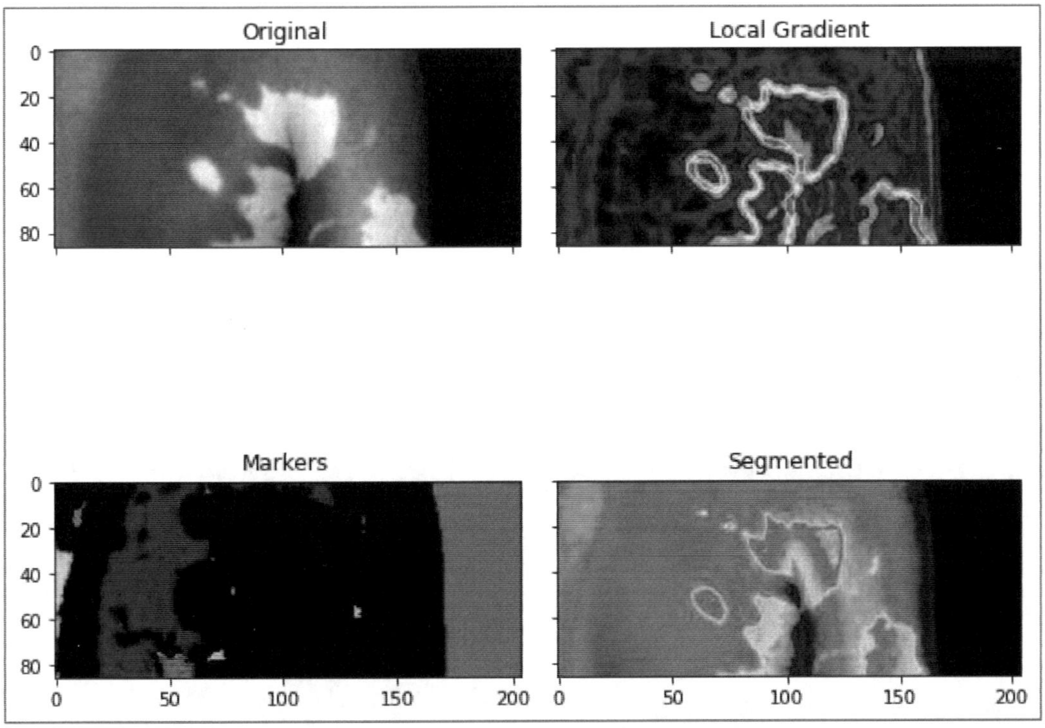

Figure 7: Steps of Watershed Algorithm.

In this method, the color image was converted into a grayscale image. The color image was segmented using the k means clustering algorithm, whereas the grayscale image was used to calculate the edges from the image. Then with the help of the initial segmented image and edge map, Region Adjacency Graph (RAG) was created. Value of connectivity was taken as 2. The result is shown in Figure 8.

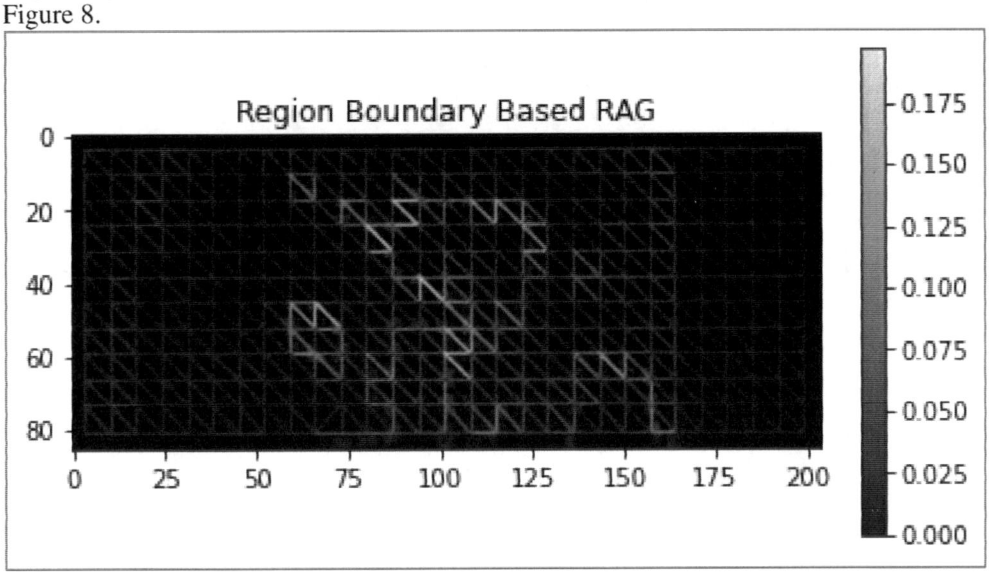

Figure 8: Result of Region Boundary Based RAG Segmentation.

In the next figure, the result of 4 segmentation techniques was shown. Felzenszwalb Segmentation was used with a value of sigma of 0.5. The number of segments decided here was 100. In Slic number of segments was taken as 250 and the value of sigma was taken as 1. Quickshift algorithm takes two important parameters kernel size which was taken as 3 and the maximum distance was taken as 6. In the watershed algorithm, a compactness feature was added for a more regularly shaped region. Finally, the result of applying all these four algorithms is shown in Figure 9.

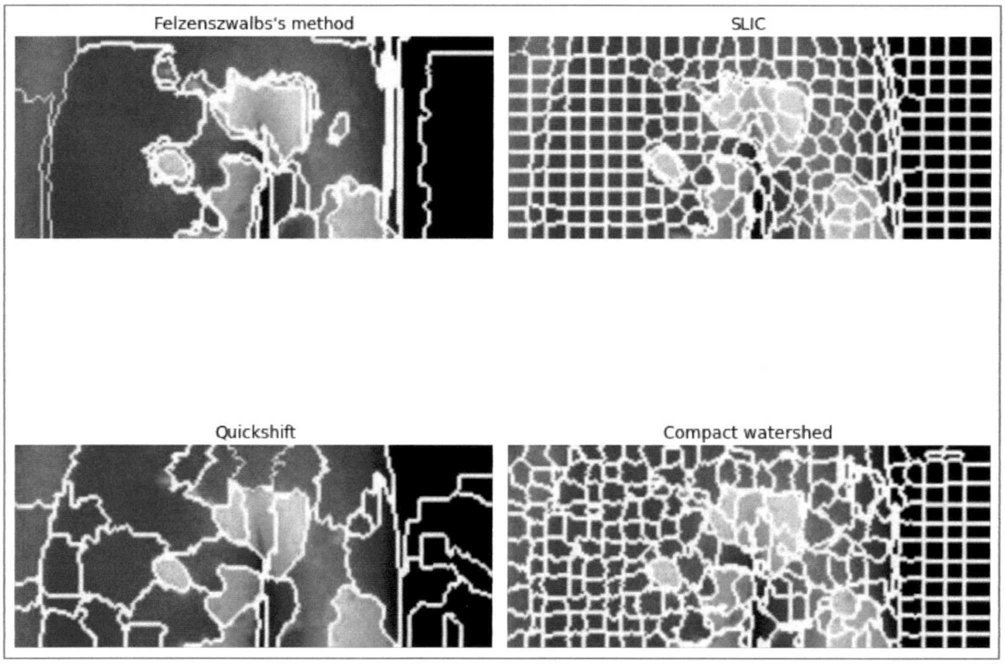

Figure 9: Comparison of Different Segmentation.

The execution time of each algorithm was measured and shown in table 1.

Table 1: Execution Time of different algorithm.

SN	Name of the algorithm	Execution time(In Millisecond)
1	Local otsu	48.7ms
2	Multi otsu	6.51 ms
3	RAG	461ms
4	Watershed	11.1ms
5	Felzenszwalb	27.9 ms
6	Slic	40.8 ms
7	Quickshift	284 ms
8	Compact Watershed	28.74 ms

5. Analysis & Discussion

As can be seen from the result and Figure 10 the normal threshold with a manual threshold is a simple and very popular technique for image segmentation but it did not work well with vitiligo images. However Otsu's thresholding gives a better result than the manual threshold but the result is still not convincing. Multi Otsu's thresholding gives the best result among all thresholding techniques. Compact watershed gives a better result than the normal watershed method. Out of Felzenszwalb, SLIC, Quickshift, and compact Watershed, Felzenszwalb, and SLIC give a promising result.

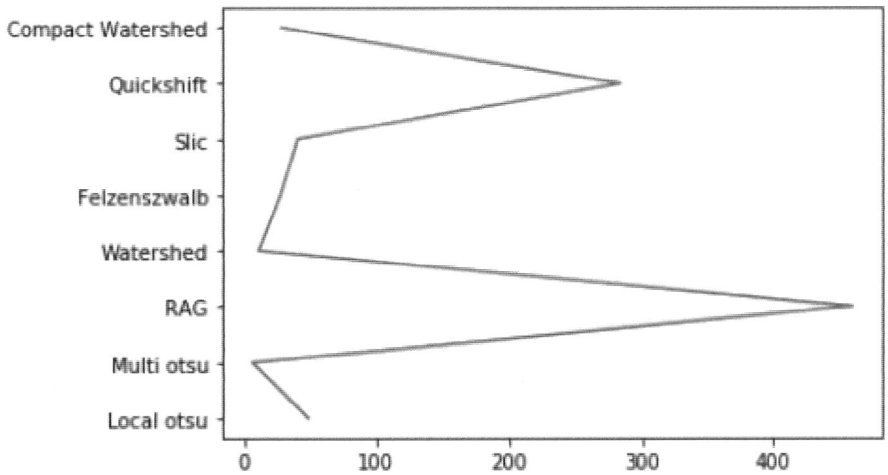

Figure 10: Comparison of Execution Time of Segmentation Algorithm

Thus by analysing results it can be seen that Felzenszwalb, SLIC, and multi Otsu's thresholding gives a better result than other techniques like Quickshift and compact watershed. In terms of performance multi Otsu thresholding gives optimal results however other overheads time (digitization) were not included in the performance. The watershed algorithm also can be considered in terms of optimal performance whereas again time is taken for denoising the image and gradient calculation was not taken into account. Felzenszwalb and compact watershed also give a comparative performance.

6. Conclusion & Future scope

This paper briefly discusses some of the segmentation techniques for vitiligo images. Image segmentation is an important step of image processing which is affected by so many features like texture, content, and various other characteristics. In this work different image segmentation techniques have been discussed and also the result of each technique on the vitiligo image has been shown. Some of the techniques which work well with other medical images do not work well with vitiligo image whereas some of the techniques give the convincing result. In the future, more research can be conducted to improve the computational speed and accuracy, as good segmentation leads to good classification accuracy.

References

[1] Liu, Jian & Yan, Jianwei & Chen, Jie & Sun, Guozhong & Luo, Wei. 2019 *Artificial Intelligence and Security* "Classification of Vitiligo Based on Convolutional Neural Network". 214-223

[2] Salamea, Christian & Chica, Juan. 2020, *Advances in Emerging Trends and Technologies* Vitiligo Detection Using Cepstral Coefficients, **1,** 389-398

[3] Pennisi, A., Bloisi, D. D., Nardi, D., Giampetruzzi, A. R., Mondino, C., & Facchiano, A. 2016 *Computerized Medical Imaging and Graphics,* Skin lesion image segmentation using Delaunay Triangulation for melanoma detection., ***52***, 89–103.

[4] Sreelatha, T., Subramanyam, M. V., & Prasad, M. N. G. 2019 *Journal of Medical Systems,* Early

Detection of Skin Cancer Using Melanoma Segmentation technique. ***43*(7)**, 1–7

[5] Anthal, J. 2017. *International Conference on Current Trends in Computer, Electrical, Electronics and Communication (CTCEEC)*, Detection of Vitiligo S kin D isease using LVQ Neural Network. *2017* 922-925

[6] Gurusamy, Vairaprakash & Kannan, Subbu & G.Nalini,. 2014 Conference: RTRICS, Review On Image Segmentation Techniques

[7] M. Low, V. Huang and P. Raina 2020 *IEEE 17th International Symposium on Biomedical Imaging (ISBI)* "Automating Vitiligo Skin Lesion Segmentation Using Convolutional Neural Networks, 2020, Iowa City, IA, USA, 2020, pp. 1-4, doi: 10.1109/ISBI45749.2020.9098682.

[8] Paolo Neri, Michela Fiaschi and Giovanni Menchini 2020,*Journal of Imaging*, Semi-Automatic Tool for Vitiligo Detection and Analysis

[9] Das, Nibaran & Pal, Anabik & Mazumder, Sanjoy & Sarkar, Somenath & Gangopadhyay, Dwijendranath & Nasipuri, Mita. 2013, *Proceedings - 2013 3rd International Conference on Advances in Computing and Communications*, ICACC 2013. 208-211. 10.1109/ICACC.2013.48. An SVM Based Skin Disease Identification Using Local Binary Patterns.

[10] Arnab Banerjeea, Nibaran Das, Mita Nasipurib 12020 Skin Diseases Detection using LBP and WLD- An Ensembling Approach, arXiv:2004.04122v1 [cs.CV] 8 Apr 2020

[11] Mahrishi, M., Hiran, K. K., Meena, G., & Sharma, P. (2020). Machine Learning and Deep Learning in Real-Time Applications. IGI Global. \

[12] Mete, M., & Sirakov, N. M. 2010. *BMC Bioinformatics* Lesion detection in demoscopy images with novel density-based and active contour approaches. 11(SUPPL. 6). https://doi.org/10.1186/1471-2105-11-23

[13] Ann Nosseir,Mokhtar Ahmed Shawky 2019*, Proceedings of the 2019 8th International Conference on Software and Information Engineering* Automatic Classifier for Skin Disease Using k-NN and SVM/ICSIEEgypt

[14] Ashmita Gupta ,Mansi Bhatnagar ,Ashish Issac ,Malay Kishore Dutta ,Carlos M Travieso 2019, *APPIS '19: Proceedings of the 2nd International Conference on Applications of Intelligent Systems* ,Imaging method for Noise Removal and segmentation of Skin Lesions from Dermoscopic Images /APPIS Spain.

IOP Conf. Series: Materials Science and Engineering 1131 (2021) 012004 doi:10.1088/1757-899X/1131/1/012004

A Comparative Study of Various Human Activity Recognition Approaches

Dhruv Goel[1] and Rahul Pradhan[2]

GLA University, Mathura

[1]dhruvgoel297@gmail.com, [2]rahul.pradhan@gla.ac.in

Abstract. Human Activity Recognition (HAR) is a vast and exciting topic for researchers and students. HAR aims to recognize activities by observing the actions of subjects and surrounding conditions. This topic also has many significant and futuristic applications and a basis of many automated tasks like 24*7 security surveillance, healthcare, laws regulations, automatic vehicle controls, game controls by human motion detection, basically human-computer interaction. This survey paper focuses on reviewing other research papers on sensing technologies used in HAR. This paper has covered distinct research in which researchers collect data from smartphones; some use a surveillance camera system to get video clips. Most of the researchers used videos to train their systems to recognize human activities collected from YouTubes and other video sources. Several sensor-based approaches have also covered in this survey paper to study and predict human activities, such as accelerometer, gyroscope, and many more. Some of the papers also used technologies like a Convolutional neural network (CNN) with spatiotemporal three-dimensional (3D) kernels for model development and then using to integrate it with OpenCV. There are also work done for Alzheimer's patient in the Healthcare sector, used for their better performance in day-to-day tasks. We will analyze the research using both classic and less commonly known classifiers on distinct datasets available on the UCI Machine Learning Repository. We describe each researcher's approaches, compare the technologies used, and conclude the adequate technology for Human Activity Recognition. Every research will be discussed in detail in this survey paper to get a brief knowledge of activity recognition.

1. Introduction

Our research topic, i.e., Human Activity Recognition (HAR) is a very vast and exciting topic for research and a mode of attraction for students. There are already many kinds of research done in this domain of Computer Vision and AI. These researches also help automate applications when implemented for real world performance, like long-distance broadcasts, video surveillance, human-computer interaction (HCI), and many more in the healthcare sector. However, it is also having its difficulty phase in terms of sensor motions and their installments; background activities also affect the model outcomes, variability in the activities played by different humans, variation in scale, viewpoint, and partial occlusion. As the thinking capabilities advance and hardware configurations for HAR system upgrades, research approaches are continually increasing. In this survey paper, much research on sensing technologies, video analysis for activities, and computer vision techniques used in HAR are reviewed. The most commonly used techniques for the HAR system, irrespective of computational powers or activities classification algorithms, are considered.

As we approach any separation process, we should consider two questions: "What action?" about the recognition problem, and "Where is the video?" with a local problem. When trying to identify human activity, one has to determine the person's kinetic state so that the computer can see the activity. The

Content from this work may be used under the terms of the Creative Commons Attribution 3.0 licence. Any further distribution of this work must maintain attribution to the author(s) and the title of the work, journal citation and DOI.
Published under licence by IOP Publishing Ltd

system can quickly determine some activities which happen in daily life, such as "walking," "running," "yoga," and many more. While in comparison with daily activities, activities like "peeling apple" are more complex to determine. They should be divided into more specific activities to get recognized by the system. The HAR system should first recognize the objects to better recognize the activity as it provides more information about the ongoing event [1].

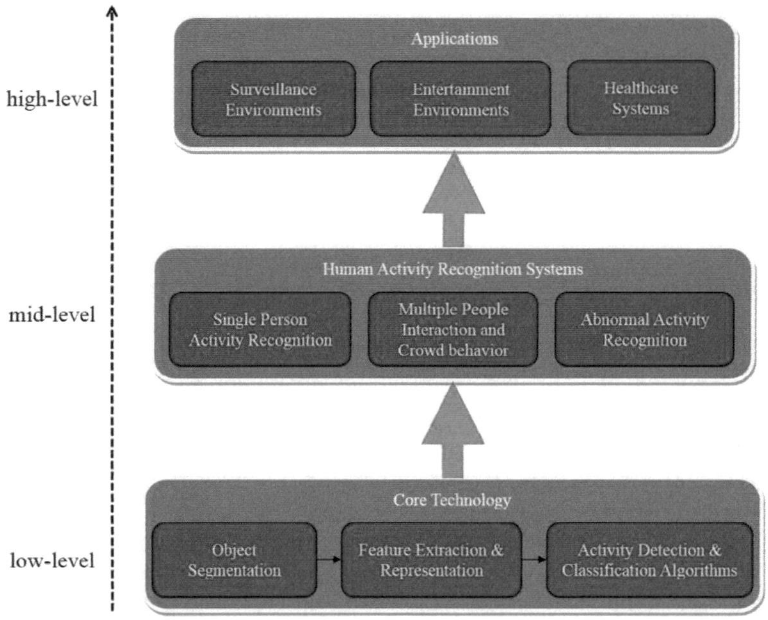

Figure 1: General Human Activity Recognition system.

HAR system has an inherent hierarchical structure, which indicates three different levels (figure 1). First, core technologies are used step by step; the first step is object segmentation followed by feature extraction from each frame of the video, and then using algorithms will help classify the human activities. The first level is then needed to integrate with some system to observe the results of algorithms and video processing in which the second layer comes into the picture. The HAR system comprises three categories of activities namely, recognition of one-on-one work, social interaction and crowd behaviour, and unusual activity recognition. Finally, the top layer, i.e., application implementation of the HAR system, can be performed in any places, majorly crowded areas for better law regulations like surveillance environments, entertainment environments, and healthcare systems.

Human activities also follow a hierarchical structure having activities divided into three categories. First is action primitive; this consists of complex activities. The second category comprises action/activities, and then finally, complex integration is a third category that refers to human activities involving people or objects over two people. In this survey paper, we will also combine these three categories. The first verb refers to activities that involve parts of the human body, such as the arms, legs, hands, upper limbs, and many more [2]. This category includes activities like "stretching the upper body," "rotating hands," and many others. The second category, i.e., Actions and activities, refers to the movement of the whole body made in the first few actions in a series of moments and performed by one person in the frame. This category includes activities such as walking, eating, upstairs, yoga, and many others. Interactions refer to the activities performed by more than one person, such as arm wrestling, and these types of activities are sometimes very complex to recognize.

Also, for real world deployment of HAR, efficient video analysis is very important, both in online and on the edge mode of video recognition. To get efficient results we need to use more advance technologies like deep learning [86, 87, 78, 88, 93, 89, 92]. Video understanding is more difficult as compare to image recognitions, as videos include temporal models. For e.g., for distinguishing between forward and backward walking, reversing the sequence will give different results, so temporal modelling is essential. Many previous approaches use 2D CNN [84, 77, 87, 92]. In recent approaches 3D is being a most promising techniques in video recognition as they consider both spatial and temporal features, but we have to compromise on computational cost and deployment, that's why it cannot be applied in online video recognition. We proposed a method in consideration of all the constraints i.e., Temporal Shift Module (TSM) (figure 2).

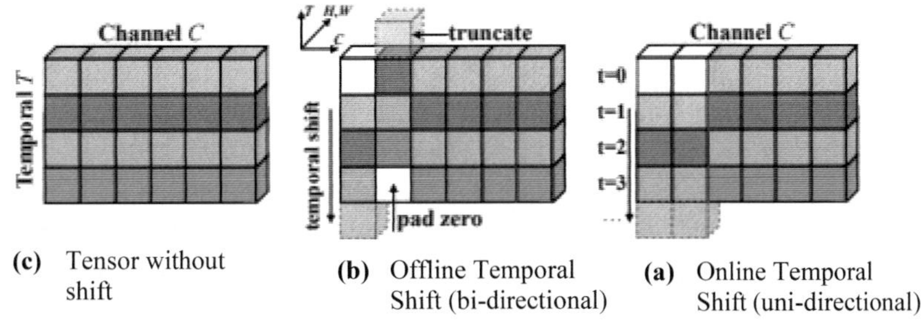

Figure 2: General representation of TSM approaches

We tried the shift strategy used in image recognition for video recognition too, but we face two major issues: (1) *not efficient*: shift operation is conceptually zero FLOP but it is having data movement. And the extra data movement cost too much which is non-negligible and will result in increase of latency. (2) *not accurate*: too many shifting of channels will lead to performance degradation as it will destroy the spatial modelling capability. To tackle with these problems, we have come up with two solutions, (1) instead of moving all the channels, we will be moving some of its portion turn by turn, which will reduce the data movement cost. (figure 3a) (2) Use of residual branch, which will preserve the activation of the current frame, and will not destroy 2D CNN spatial modelling capability.

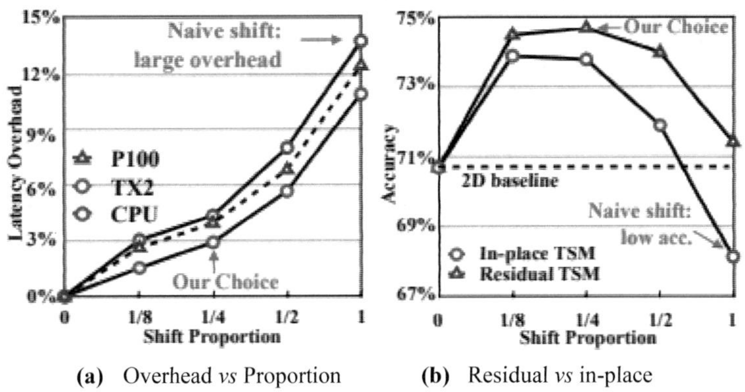

Figure 3: (a) Latency due to data movement. (b) Accuracy variation by shifting approach

Many researchers have already concluded the HAR system in three ways of representation, i.e., Global, Local, and Depth-based. Researchers of [3] use image descriptors as space-time shapes, a clear example of Global representation. The evolution comes with space-time interest point (STIP), proposed in [4], an example of local representation that uses HOG (Histogram of Oriented Gradients) and HOF (Histogram of optical flow) as the image descriptors for informative interest points. Depth-based representation becomes a new research topic with advancements in cameras with an RGB camera, different sensors like accelerometer, gyroscope, and many others.

Machine learning classification techniques like DTW (Dynamic Time Warping) were designed for speech recognition; also Hidden Markov Model (HMM) was one of them; these algorithms were not explicitly designed for HAR [5, 6]. Also, many deep learning techniques were developed for a large amount of image classification [7]. With the increase in crime rates, another research area is proposed, i.e., Human tracking to keep surveillance over abnormal behaviours.

The rest of the paper covers the following sections: in section 2, we will discuss the researches covered in some previous surveys, then further in section 3, we will cover some common challenges faced in human activity recognition and how to overcome them. In section 4, we presented human activities' categorization as local and global representations in RGB videos and depth image approaches. Section 5 reviewed the various HAR methods and classification techniques and analyzed the pros and cons of different approaches used for HAR. In section 6, we reviewed different human tracking methods and discussed some futuristic research topics in HAR. Finally, at last, we will conclude our survey in the last section.

2. Previous Surveys and Taxonomy

Numerous research published are covered area of human activity recognition. [8] divided the study into 2D (with exact and external shapes) and 3D methods. In [9], introduced a new tax code focusing on mobility, followed by a deep sensors multiview cameras, and the acquisition of activities recognition. In line with the spirit of the previous tax [10] the successive category of management has been promoted. A study by [11] focused on high-performance perceptions and raised four times as much tax, including the beginning of the human movement, tracking, positioning, and visualization.

[12] proposed research differentiating between "actions" and "activities," in this research author categorizes activities according to their activity's complexities. [13] also categorize human activity methods in their approach in two categories, i.e., "top-down" and "bottom-up." In the same way, [14] also divided the methods of human activity recognition into tree-based taxonomy into two categories, "single layer" and "hierarchical" methods. Researchers like the proposed methods [15, 16] by modeling 3D data. Work recognition with representation of the human body in 3D modelling is more accurate and beneficial for analysis than 2D functions seen in photographs, as deep cameras are used to detect the most powerful features of connected organs and joints in the human body. [17] The recently proposed study focuses on methods that exploit 3D depth data from two phases of HAR 3D stereo systems and motion capture systems. Motion imaging systems have also played their role in the skeletal muscles described using deep sensors.

[18] suggested the HAR method from standing images as all known methods were used for video processing, and most HAR systems are proposed to be known for face and measurement techniques. As [19] proposed computer-to-human communication [52,53], the focus was on facial expressions, placement measurement process, and speech recognition. Research is proposed on state recognition methods that include non-verbal means, such as facial expressions and facial expressions. [21] have studied modern HAR techniques, and have gone through many open computer problems and how they can enhance the efficiency of the human-computer system. [22] presented a review of modern observational perspectives using critical and perceptual autonomous states and provided a list of data-related data ideas related to human speech. [23] suggested analyzing a variety of non-verbal forms (i.e., visual and auditory cues) behavioural recognition patterns and data details of automated agreements and disagreements. Such social values can play an important role in analyzing social behaviour, which is the key to public participation. Finally, a complete technical analysis of human behaviour recognition from a data perspective and information presented in [24].

2D CNN methods with two types of streamed data i.e., spatial and temporal are more efficient than 3D approaches [77,91], as Temporal Segment Networks [87] use some segments of frames to extract features. But this is not much effective in more complex temporal frames and videos. 3D CNN can efficiently learn spatial-temporal frames. As they are costly in terms of computation and deployment is also difficult. 3D CNN is also having more hyper parameters than 2D CNNs, that make it more chances for over-fitting. Our TSM approach is also just like 3D CNN with spatial-temporal modelling ability and computational abilities of 2D CNNs. As discussed above, we can use 3D CNN method for direct temporal modelling. An alternative way to perform temporal modelling is with the help of 2D CNN and post-hoc fusion [83, 80, 92, 79]. Use of LSTM to aggregate with 2D CNN features also proposed in previous studies [90, 79, 85, 81, 82].

Transfer Learning is used to transfer useful information for better classification like weights of the model [94], features, objects of the classifiers. A most in-demand application of transfer learning is used in CNN to predict the diseases by classification from images [95, 96]. Transfer learning is also used in transferring images features to ECG machine for disease prediction [97]. Among all of these applications and methods one important thing is to know what parameters to transfer prior implementing final model.

3. Challenges in Human Activity Recognition (HAR)

3.1 Intraclass Variation and Interclass Similarity

There is no grammar and a firm definition of human activity, unlike speech recognition. This creates double confusion. On the other hand, the same function can vary from topic to topic, leading to intraclass diversity. Play, speed and energy also increase meeting spaces. On the other hand, different tasks may present similar situations (e.g., using laptops and reading). These functions are called interclass simulations, which are more common in HAR. Accurate and differentiating features need to be developed and removed from activity videos to address these issues.

3.2 Recognition under Real-World Settings

3.2.1 *Unsaturated Backgrounds*. Performance recognition's complexity comes when most situations use dynamic recording devices, while applications such as video recognition and fall system use still cameras. A typical example of a powerful recording is a sports event broadcast. People uses portable devices with embedded cameras and smart glasses also to record the streaming video. These streaming videos having more complex background with changing location which make them strong and complex backgrounds to recognise the activity. Second, the actual videos are full of visuals, lighting variations, and viewing changes, making it challenging to see tasks in complex and different situations

3.2.2 *More than two subject Interactions and Crowded Activities*. Previous research focused on human activities of low-level types such as running, jumping, and raising hands. One common feature of these functions is based on single type communication without personal or human communication. They involve only an individual in an activity. However, people often do activities with one or more people and objects in the real world. The football game is an excellent example of a partnership where many players (i.e., personal contact) in a team defend the ball (e.g., personal contact) and compete with players from another team. It is a challenge to find and follow many topics in a row or to see all the team's activities as "playing football" instead of "it works."

(a) (b)

Figure 4: Long-Distance and Low-Quality videos: (a) Abnormal behaviours in surveillance. (b) HAR in long-distance broadcasts.

3.2.3 *Distanced and Degraded-Quality Videos*. Far distanced and low-impact videos with high visibility are available in most video viewing contexts. Larger and denser areas such as airport municipalities and passenger passengers are independent times where things happen more often. Besides, high-quality surveillance cameras cannot deliver high-quality videos like current data sets when the target person is clear and transparent. While we do not expect to follow everyone in these situations, some unusual or criminal-related behaviour should be identified by the HAR system (Figure 4 (a)). Another common long-distance case is the distribution of the ball (Figure 4 (b)). Due to the long-range of cameras, the issue is small, making it difficult to analyze the torso [10], and the low quality of those long-distance videos increases the difficulty.

3.2.4 *How to overcome these challenges*. To overcome these challenges, we need to include alternatives, namely, (i) backlash [25, 26]; in this case, the system distinguishes a domain that does not function over time (background) in relation to people, moving or changing objects in the framework (front of the earth); (ii) tracking a person, in this case the tracking of a person's position over time, which also helps to follow an unusual behaviour [27, 28, 29]; and (iii) the discovery of a person's object and the acquisition of an act [30, 31, 32], in this system can identify the activities and actions of the person in the image.

4. Human Activity Representation and Categorization

4.1 Depth-Based Representation

Previous HAR researches takes a look at traditional RGB cameras to captured feasible quality videos. Whereas Deep cameras are having complex performance and high cost that's why they are limited for researches [34]. Some less expensive cameras such as Microsoft Kinetics [35] are present which helps researchers a lot with similar quality of data and depth resolutions and also, they are more accessible and more comfortable. To detect joints in real time scenarios there are kinetic SDK (accept algorithms in [35]). There's a major impact on computer viewing community due to availability of depth maps and skeletal information. These factors are prominent solution for solving HAR problem i.e., instead of conventional RGB-based methods use of in-depth maps or also can be used as enhancers in RGB designed approaches. This section has reviewed recent developments in job representation separately using in-depth maps or skeletons.

4.1.1 *Depth Maps Representation*. In-depth maps contain deeper links compared to standard color images and are very informative. The methods presented in this section view depth maps as spacecraft and extract features directly from them. These features are used independently or integrated with the RGB channel to create multiple features. Li et al. [36] used an action graph model that represents tasks that use many key positions that serve as locations in the action graph. All functions share the same standing sets, and each stand looks like a 3D dot bag from a depth map. However, incorporating all 3D points is expensive for a computer; therefore, proposing a more direct and effective method of sampling 3D representation points, achieving an accuracy of more than 90% recognition by sampling approximately 1% point report.

4.1.2 *Skeleton-Based Representations*. Skeletal structure of Bones and integrated joint's positions are the main focused features produced in deep maps. The Kinect device is ubiquitous in this case due to easy access to bones and joints. The Kinect v1 SDK system has 20 members, while the latest version (Kinect v2) has 25 members, adding five members close to the hands and neck (see Figure 4). We have reviewed recent papers on skeleton-based presentations and summarized three aspects of efforts to improve the effectiveness of skeleton-based presentations.

First, the skeletal model has a natural defect that is always a problem with the skeletal system when working at the edges (see Figure 5) [36]. Factors from the bones and spine can end up being 1wrong. Current methods often solve it by combining other solid materials to close or reduce the problems of closure by separating all the bones in different parts of the body and behaving independently because not all body parts are included.

4.2 Human Activity Categorization
The categorization of human activity recognition tasks is the most challenging task as they involve classification among different datasets that the model gets to train and test the system. Many researchers have divided HAR programs into two main categories, i. e., (i) Unimodal approaches and (ii) multidisciplinary approaches. After that further division, the two categories were divided into some more subdivisions.

Unimodal approaches detect activities from single-dimensional data, such as images, and are subdivided into (i) space-time, (ii) stochastic, (iii) statutory-based, (iv) structural-based methods.

As we discuss space approaches, transient factors include job recognition [37, 38] or trajectories [39, 40]. Stochastic methods use a mathematical model to identify human activities (e.g., Markov's hidden models) [41, 42] and, as the name suggests in methods based on the Rule, use rules to identify human activities [43, 44]. The most effective approach is modeling, using high-level thinking to mimic body movements [45, 46].

Multimodal approaches encompass all the collected data from different sources [47] and are then divided into three categories: (i) relevant, (ii) behavioural, and (iii) human communication methods.

Affected methods use communication methods to determine a person's feelings and state of touch [48, 49]. Behavioural approaches focus on facial expressions, hand gestures, and placement of recognition to determine a person's work [50, 51].

5. Activity Classification Methods and Approaches

5.1 Unimodal Methods
Unimodal methods use single modality data, such as images. Many methods present task recognition as a set of visual objects extracted from videos or images and use multiple classification models to detect this function [54, 55]. Some methods use captions of movement trajectories as it will be challenging to ensure that the movement is continuous or not [56, 57], while others use the light flow feature, which helps them see activities from full-motion video.

5.1.1 *Space-Time Methods*. The excessive amount of work known is based on the space system [79]. It is understandable that the timing approach is based on human movements' movement over time and is used in human tracking systems to monitor abnormal behaviour. This tracking space method is based on positions, footprints and travel descriptions (Wang et al., 2013). (Efros et al. 2003) presented a study on long-distance recognizing where objects have small resolutions. Numerous such research are presenting different applications of space-time methods.

5.1.2 *Stochastic Methods*. The stochastic approach is the most commonly used approach from the beginning of the research area in computer vision and Machine Learning. Many researchers presented the use of Hidden Markov Model, many researchers also consider Hidden Conditional Random Fields for stochastic models (Bishop 2006, Quattoni et al. 2007). Human behaviour detection forms sequences of actions; this is presented in research by (Robertson and Reid et al. (2006), providing information vectors of positions, velocity, and local descriptors. These features are then used by the HMM model to predict human activities.

5.1.3 *Rule-Based Methods*. This method used pre-set sets of instructions to perform the recognition. Each sets of instruction is integrated used for development of a proper efficient human activity recognising model. Many types of research are conducted using a rule-based approach in HAR, and the rule-based approach is less regulated than others, as there are specific rules set out in the system.

5.1.4 *Shape-Based Methods*. Shape-based methods are all about human pose detection and body parts' appearance to keep track of human activities. Human emotions are represented as 2D space as rectangular patches, and the volume of motions are represented as 3D space as volumetric shapes. Human bodies are presented as limbs jointly connected. By fetching all this information from videos, algorithms can detect the activities more clearly and accurately.

5.2 Multimodal Methods

In recent years, much attention has been given to the identification of multiple jobs. Many approaches are based on two different type of integration: initial integration and time integration. One of the easiest ways to get efficient performance is to integrate features in a wide vector and get the result of feature extraction (Sun et al., 2009). This combination method can improve visual performance, but the vector of the new feature is much larger.

Multimodal indicators are often compiled over time; therefore, the temporary organization of a large event and various approaches is important to understand the data. Other than visual synchronization, audio-visual analysis is used for many other purposes like keeping an eye on human trajectories (Lichtenauer et al., 2011, Perez et al., 2004) and active identity (Wu et al., 2013). Multimodal approaches are considered as three approaches: (i) affective modes, (ii) ethical approaches, and (iii) modes based on social media interaction. These methods describe the action of an atom or a combination that may be related to the emotional state of the person he is talking to emotionally and physically.

5.2.1 *Affective Methods*. This method is related to a person's emotional state, which can be recognized by hand gestures, speech, facial expressions, psychological changes, and activities. This is a combined form of computer vision, artificial intelligence, pattern recognition, and emotional science. This can be used to advance HAR by recognizing more proper and formal activities like "ongoing meeting" instead of predicting some people are sitting and talking, "a man trying to catch a bus" instead of running. Researches [20] are all about this approach of combining the emotional state of a person with its activity prediction.

5.2.2 *Behavioural Methods*. This method is very similar to the affective method as in this, we will be talking about a person's behaviour, and in the previous method, we were talking about whether a person is affected or not overtime, which can be viewed by a change in a person's psychological state. Factors that can affect a person's mood or behaviour can be segregated into different features like mood, emotions, interactions with other humans, actions, and activity. Hence, it will be difficult to predict the complex actions and human behaviour. Research in [70] presented an advancement for behaviour detection by integrating speech and visual detection.

5.2.3 *Methods Based on Social Networking*. A very basic task in every human life is social interaction, this is also a best way of detecting human activity as how frequent a person post their daily updates anyone can get the track of their activities and also there can be misuse of this in kidnapping or any other abnormal activity. Most common platforms for social interactions like Facebook, Youtube, and twitter measure social involvement of any person. A recent survey [71] provide full up-to-date information to automate human behaviour analysis.

5.2.4 *Multimodal feature fusion*. In this multimodal feature fusion, we try to recognize people's activities in a crowded area where everyone is performing different activities with different sounds. In this case, the HAR system will use visual information to recognize underlying activities easily recognized. However, the HAR system will perform recognition more accurately by taking account of audio-visual analysis. In crowded areas, many humans perform similar actions with their body parts but emit different audio. Through this audio-visual analysis, the HAR system can easily recognize the activities. The audio-visual information will help the system understand who the target person is for test sequences and distinguish different activities performed by different persons. The dimensionality of data from different modalities collected is the biggest challenge of multimodal recognition. As the dimension of video, data is much more complicated than that of audio data, so in this case, dimensionality reduction is useful. To perform these two types of techniques, we can use that proposed in [72, 73].

Early Fusion, is the process that occur on extraction level to extract features from different frames, at the time of extracting features and combining different modality by dimensionality reduction of data and forming a single vector that is viewed as an individual. The main advantage is that when we have so many different modalities that are positively correlated, the result will be more accurate since, in this Fusion, we need only a single learning phase. However, this approach may fail at the time of one modality dominating over others.

Another category is late Fusion; this Fusion is done at the decision level. It learns parameters of each modality separately by several probabilistic models. They will be combined in a supervised framework providing a final score [75, 76]. The strengths of each modality affect the results of recognition in a better way. However, this approach is time-consuming in comparison to another approach.

Also, a third approach is presenting by [74], known as slow Fusion. This approach combines the above-discussed approaches and acts as a hierarchical approach, this will pass all the layers from a fusion layer to get combined data and also this process is slightly a slower process. Furthermore, it is having advantages of both the fusion levels, but there is also a computational restriction because of the different level of raw data that is to be processed.

5.3 Classification Approaches

As the next stage, we will discuss some classification algorithms to recognize human activities from features extracted from videos and images. The classification approaches are considered in mainly three categories: frame pattern matching, discriminative models and generative models. Pattern matching-based methods are relatively more straightforward but sometimes expensive because of its computational requirements. Generative algorithms learn using joint probabilities. Now comes with discriminative methods that give results directly as generative methods include hidden Markov model, dynamic bayesian method, and discriminative approach include models such as Artificial neural network (ANN), relevance vector machine (RVM) and support vector machines (SVM).

5.3.1 Template-Based Approaches.

5.3.1.1 *Template Matching.* The template matching approach uses two components to match on shapetime, i.e., the motion-energy image that captures any motion over time, motion-history image that captures the last occurrence of any motion. Furthermore, when found in the frame, it generates a template of an activity. They can also be regarded as a shape-time approach. This approach is presented in [68, 69].

5.3.1.2 *Dynamic Time Warping.* This method is just like its name, a dynamic programming algorithm initially developed for speech recognition which is used as matching two sequences with variances. Its task to represent the words as templates and if found a new word than giving a word score for it. When it comes to applying DTW in HAR since human activities captured videos and videos that are collected online too is considered as a sequence of keyframes. Then the activity recognition task is then transformed to a frame pattern matching problem. Then research [67] proposed the solution of learning gestures using pre-trained models. The result was gained after gestures features after matching patterns among frames can be obtained from means and variances of correlations between features of image frames and new videos.

5.3.2 Generative Models

5.3.2.1 *Hidden Markov Model Approach.* One of the tasks of hidden markov model is human activity recognition. Hidden Markov Model is first proposed for speech recognition. In research [63], Yamato et al. was the first one to use HMM for activity recognition. They used the number of pixels as feature extracted from each frame. Then they use HMM to train on features combined vector of sequence from each frame including some probability gained by some matrices, confusion matrix and transition matrix. But after all this, the problem gets transformed into a simple evolution problem. Then some advancements are presented to combine human body parts and track them for feature extraction. Then in combination of all the approaches, researchers presented an approach which is a coupled HMM approach used for human computer interaction in research [64-66].

5.3.2.2 *Dynamic Bayesian Networks.* A dynamic bayesian network, unlike HMMs, has an essential feature as it contains more than one random variable, which can contain only one random variable. In this way, HMMs are the simplified form of DBN with a restricted number of the random variable and a fixed graph structure. Research [62] proposed a model for two hands gestures; in this, we can see three hidden variables. These three hidden variables represent two hand motions and their spatial relation. Furthermore, we can extract at least five features, i.e., two hands motion and relative position with face and spatial relation observed between two hands.

5.3.3 Discriminative Models

5.3.3.1 *Support Vector Machines.* This method is the most commonly used model applicable in real world for discriminative models, and now getting started in HAR. Preciously, research [60] presented an SVM model to distinguish between two classes. In this, the author proposed a model to create a margin between two classes. Furthermore, research [61] presented a combination of space-time features and combined their SVM model to implement the HAR system through video processing. A dataset known as KTH sets a benchmark in HAR research areas; they used this data for their model training.

5.3.3.2 *Conditional Random Fields.* This method compactly represents the conditional probability of an event to be held, i.e., the sequence of event Y from the combinations of sequence X, and it is a type of undirected graphical model. [59] compared HMMs and CRFs, and they discovered that CRF's performance is better than HMMs, although HMM's model features are independent of the assumptions of HMMs. Whereas the HMMs model is independent of the given labels; thus, these complex features invalidate its assumptions, and the method that is no longer a proper generative model is making of HMM.

5.3.3.3 *Deep Learning Architectures.* In the deep learning approach, we are having four approaches, i.e., Recurrent neural network (RNNs), Deep neural networks (DNNs), convolutional neural network (CNNs), and some emergent neural networks. Among these categories, ConVents is the one which is widely used for prediction with deep learning architecture. However, in HAR's case, the dataset that we get is very small compared to the ideal data that Convent needed to be trained on. One research of [58] represents the solution to combining hand-crafted features with improving the performance.

5.3.4 Temporal Shift Module

5.3.4.1 *Offline Video Recognition.* After we get videos, we first grouped T frames as a sample to process on 2D CNN models. The model processed each frame and test them on a threshold and gathered all the results for final prediction. Our TSM model performs on same computational cost as 2D CNN performs. TSM will work on each frame's residual block and this help in better temporal fusion of information at low computation. TSM is also beneficial in conversion of any 2D CNN model into 3D CNN model and this will not require any extra computation. This make our TSM model compatible on any hardware specifications, just with a pre-requisite of 2D CNN compatibility.

5.3.4.2 *Online Video Recognition.* Online video recognition is most important in real-life deployment of HAR. This is sometime difficult task to get a solution with high accuracies and low time complexities. Online video recognition can be used in Augmented Reality or Self driving cars for crowd detection and their activities and also in its complex applications. Our TSM model is capable of achieving acceptable accuracies for online video recognition.

In Offline mode, TSM uses features of succeeding frames to use in the current frame that's why it is also known as bi-directional shifts (figure 2). In online mode TSM performs smart functions as it uses preceding frame feature for current frames, in this way our model performs shift in uni-direction. The online video recognition gives us low latency as we use only a minor part of frames to get features whereas some methods [93] uses large number of frames to predict one result which leads to high time-complexity. Low memory consumption will be there as we are saving only that minor part of frames in cache memory.

5.3.5 *Transfer Learning.* In HAR transfer learning is emerging because of many important applications such as disorder of movement, obesity, and patient monitoring from distance in some deadly diseases [98, 99]. This method starts with clustering of videos as much as possible to segregate similar actions and to predict them in clusters. Then, the analyzing of CNN architecture layers will be done by using CCA similarity [100]. From the report generated we will be able to know what features to transfer and tune the layers again for more accuracy.

5.3.6 *BERT in 3D CNN architecture*

5.3.6.1 *Temporal Modelling.* In this method, researchers first perform classification with the help of 3D CNN excluding the BERT based temporal model and they extracted features from the output of classification. Then to preserves the information they use BERT classification approach by adding those extracted features to the parameters of BERT encoding and also additional classification features appended in the results [101].

5.3.6.2 *Feature Reduction.* BERT is producing too many parameters to be worked on. As if there are approx. 500 feature dimensions then the BERT will have approx. 2-3 million of parameters and if we increase the features dimension as this will depend on the inputs, BERT parameters will increase immensely. There is the need to reduce features for this, researchers come up with two approaches i.e., with help of modified block which will be used for feature reduction (FRMB) and also an approach of feature reduction with additional block (FRAB) is also proposed along 3D CNN model backbone model. In FRMB, the backbone model gets replaced with a block which will be reducing dimensions of features. While in FRAB, they use an additional block for features dimension reduction (for e.g., ResNetXt101 as backbone of 3D CNN architecture figure *5).*

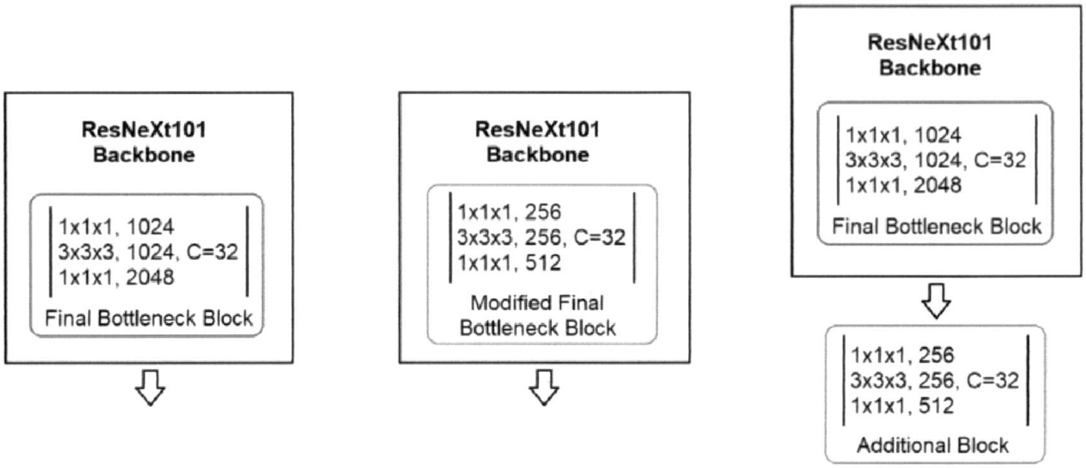

(c) Original **(b)** FRMB **(a)** FRAB

Figure 5: Feature Reduction Approaches

5.3.6.3 *SlowFast Architecture.* SlowFast approach [102] includes both streams fast and slow. As before this approach there were different modalities for each stream, but now there will be only one architecture that will work for both streams namely fast and slow. Now, due to its different streams lead to different temporal resolutions will not allow BERT temporal approach to implement. In this, two different approaches are proposed Early Fusion and Late Fusion. In early fusion, only one BERT model is utilized, and the concatenation of streams will be done before the BERT layer. And to make this merger efficient there will be reduction of temporal resolution for fast stream. In late fusion, the outputs of two BERT layers from each stream will be concatenated as we need two different BERT models in this fusion (figure 6).

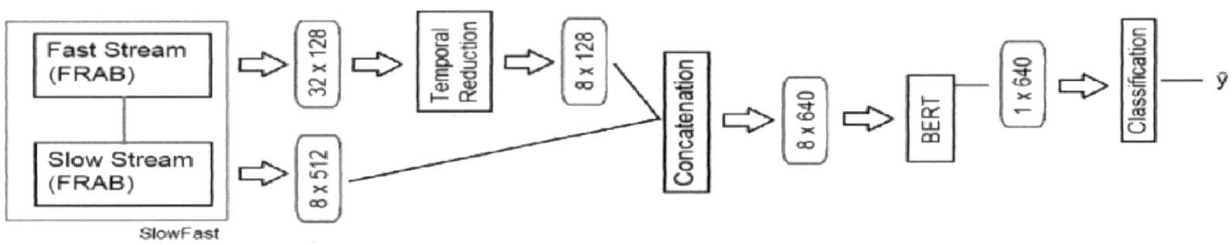

(b) Early Fusion

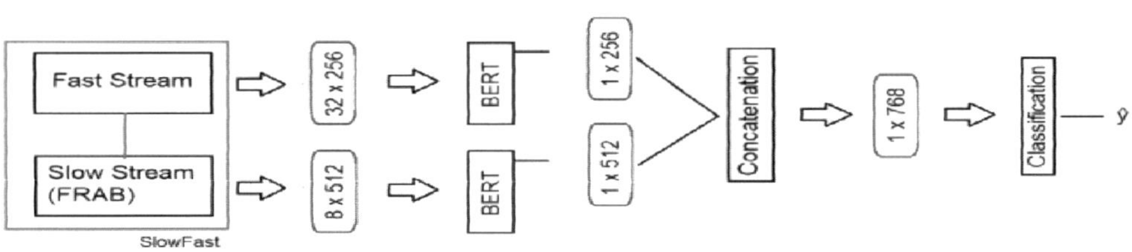

(a) Late Fusion

Figure 6: Implementation of both Fusion techniques

6. Human Tracking approaches and Futuristic approaches

6.1 Human Tracking

Another topic of concern for research is human tracking for video surveillance. In video surveillance this process will locate human's trajectories, and then the system analyzes the person's trajectories for better performance. This tracking of human trajectories will help in abnormal behaviours. Human tracking [33] can be performed using two approaches, i.e., (i) kernel-based approach and (ii) filtering-based approach.

6.1.1 *Filter Based Tracking*. In the filter-based method, there are two most used techniques, i.e., Particle filter (PF) and Kalman filter (KF), and also filter based approach is itself a widely used technique to track the humans. Gaussian noise perturbed a state estimation method i.e., KF and it is based on linear dynamical systems. Furthermore, for non-gaussian and multimodality cases, we have particle filter, it is a conditional density propagation method.

6.1.2 *Kernel Based Tracking*. In the kernel-based tracking, the system uses color histograms of weighted objects to track the objects from one frame to another over time, and this process is iteratively performed on mean shift procedure is also called mean-shift tracking. The mean shift procedure provides the fast speed of convergence and low requirements of computation.

6.2 Further Discussion

In this section, we will be discussing three futuristic solutions, applications, and advancements.

First, the biggest challenge in HAR is computational power as we have a computational constrained system, and to overcome this, we can go for embedded sensors, microchips in hardware. Furthermore, we can go for more advanced classification approaches to recognize activity in patches from a computer vision perspective. We can also degrade the quality of the input image and focus on balancing algorithm efficiency, information, and accuracy rate.

Second, many tasks in HAR is done on single datasets, but we can go for evolution to combine many large datasets and make complex and single data. This evolution will make our results more efficient, and also it will require advanced algorithms to train the system. Another evolution is to compare the algorithms and unite them to make a more successful model, and it will also work for real world inputs.

Third, we can make our results more attractive by including the behaviour of the surroundings like instead of saying a person is running, we should rather say a person is trying to catch the bus; instead of saying many people sitting and talking, we should rather say a meeting is going on. We should include Natural Language Processing (NLP), which will also provide us with behavioural context with actions and activities.

7. Conclusion

Human activity recognition is a very prominent topic and widely used in video surveillance, healthcare, and human-computer interaction. As technology advances, crime increases, so we need to advance our HAR application for better law regulation. Furthermore, challenges will increase when we come to real scenarios.

In this review, human activities are divided into three categories of different forms of performing actions, first action primitives, then actions/activities, and thirdly interactions. We have discussed the representation of human activity recognition, classification approaches, and the category of different approaches of HAR. We have also summarized the classification approaches, and we have taken from the beginning to the latest technology, i.e., deep learning approaches. We have gone through an advanced application of HAR, i.e., human tracking.

As there are many surveys on human activity recognition, none of them adequately describe human activity recognition as they discuss the computational issues in human activity recognition. At last, we can conclude that, though humans keep on understanding different methods for HAR, there are more severe challenges in real scenarios such as human poses detection, handling background errors, occlusion, and many more.

References

[1] Gupta, A., and Davis, L. S. (2007). "Objects in action: an approach for combining action understanding and object perception," in Proc. IEEE Computer Society Conference on Computer Vision and Pattern Recognition (Minneapolis, MN), 1–8.
[2] J. M. Chaquet, E. J. Carmona, and A. Fernández-Caballero, "A survey of video datasets for human action and activity recognition," Computer Vision and Image Understanding, vol. 117, pp. 633–659, 2013
[3] M. Blank, L. Gorelick, E. Shechtman, M. Irani, and R. Basri, "Actions as space-time shapes," in Tenth IEEE International Conference on Computer Vision (ICCV'05) Volume 1, pp. 1395–1402, Beijing, China, 2005
[4] I. Laptev and T. Lindeberg, "Space-time interest points," in Proceedings Ninth IEEE International Conference on Computer Vision, vol. 1, pp. 432–439, Nice, France, 2003.
[5] L. Rabiner and B. H. Juang, Fundamentals of Speech Recognition, 1993
[6] L. R. Rabiner and B. H. Juang, "An introduction to hidden Markov models," IEEE ASSP Magazine, vol. 3, pp. 4–16, Prentice Hall, Upper Saddle River, New Jersey, 1986
[7] A. Krizhevsky, I. Sutskever, and G. E. Hinton,"Imagenet classification with deep convolutional neural networks," in Advances in Neural Information Processing Systems, pp. 1097–1105, Lake Tahoe, Nevada, 2012
[8] Gavrila, D. M. (1999). The visual analysis of human movement: a survey. Comput. Vis. Image Understand. 73, 82–98. doi:10.1006/cviu.1998.0716
[9] Aggarwal, J. K., and Cai, Q. (1999). Human motion analysis: a review. Comput. Vis. Image Understand. 73, 428–440. doi:10.1006/cviu.1998.0744
[10] Wang, L., Hu, W., and Tan, T. (2003). Recent developments in human motion analysis. Pattern Recognit. 36, 585–601. doi:10.1016/S0031-3203(02)00100-0
[11] Moeslund, T. B., Hilton, A., and Krüger, V. (2006). A survey of advances in visionbased human motion capture and analysis. Comput. Vis. Image Understand. 104, 90–126. doi:10.1016/j.cviu.2006.08.002
[12] Turaga, P. K., Chellappa, R., Subrahmanian, V. S., and Udrea, O. (2008). Machine recognition of human activities: a survey. Proc. IEEE Trans. Circuits Syst. Video Technol. 18, 1473–1488. doi:10.1109/TCSVT.2008.2005594
[13] Poppe, R. (2010). A survey on vision-based human action recognition. Image Vis. Comput. 28, 976–990. doi:10.1016/j.imavis.2009.11.014
[14] Aggarwal, J. K., and Ryoo, M. S. (2011). Human activity analysis: a review. ACM Comput. Surv. 43, 1–43. doi:10.1145/1922649.1922653
[15] Chen, L., Wei, H., and Ferryman, J. (2013b). A survey of human motion analysis using depth imagery. Pattern Recognit. Lett. 34, 1995–2006. doi:10.1016/j.patrec. 2013.02.006
[16] Ye, M., Zhang, Q., Wang, L., Zhu, J., Yangg, R., and Gall, J. (2013). "A survey on human motion analysis from depth data," in Time-of-Flight and Depth Imaging, Lecture Notes in Computer Science, Vol. 8200. eds M. Grzegorzek, C. Theobalt, R. Koch, and A. Kolb (Berlin Heidelberg: Springer), 149–187.
[17] Aggarwal, J. K., and Xia, L. (2014). Human activity recognition from 3D data: a review. Pattern Recognit. Lett. 48, 70–80. doi:10.1016/j.patrec.2014.04.011
[18] Guo, G., and Lai, A. (2014). A survey on still image based human action recognition. Pattern Recognit. 47, 3343–3361. doi:10.1016/j.patcog.2014.04.018
[19] Jaimes, A., and Sebe, N. (2007). "Multimodal human-computer interaction: a survey," in Computer Vision and Image Understanding, Vol. 108 (Special Issue on Vision for Human-Computer Interaction), 116–134.
[20] Pantic, M., and Rothkrantz, L. (2003). "Towards an affect-sensitive multimodal human-computer interaction," in Proc. IEEE, Special Issue on Multimodal Human-Computer Interaction, Invited Paper, Vol. 91 (IEEE), 1370–1390.
[21] Pantic, M., Pentland, A., Nijholt, A., and Huang, T. (2006). "Human computing and machine understanding of human behaviour: a survey," in Proc. International Conference on Multimodal Interfaces (New York, NY), 239–248

[22] Zeng, Z., Pantic, M., Roisman, G. I., and Huang, T. S. (2009). A survey of affect recognition methods: audio, visual, and spontaneous expressions. IEEE Trans. Pattern Anal. Mach. Intell. 31, 39–58. doi:10.1109/TPAMI.2008.52

[23] Bousmalis, K., Mehu, M., and Pantic, M. (2013a). Towards the automatic detection of spontaneous agreement and disagreement based on nonverbal behaviour: a survey of related cues, databases, and tools. Image Vis. Comput. 31, 203–221. doi:10.1016/j.imavis.2012.07.003

[24] Rodríguez, N. D., Cuéllar, M. P., Lilius, J., and Calvo-Flores, M. D. (2014). A survey on ontologies for human behaviour recognition. ACM Comput. Surv. 46, 1–33. doi:10.1145/2523819

[25] Elgammal, A., Duraiswami, R., Harwood, D., and Davis, L. S. (2002). Background and foreground modeling using nonparametric kernel density for visual surveillance. Proc. IEEE 90, 1151–1163. doi:10.1109/JPROC.2002.801448

[26] Mumtaz, A., Zhang, W., and Chan, A. B. (2014). "Joint motion segmentation and background estimation in dynamic scenes," in Proc. IEEE Computer Society Conference on Computer Vision and Pattern Recognition (Columbus, OH), 368–375.

[27] Liu, J., Yan, J., Tong, M., and Liu, Y. (2010). "A Bayesian framework for 3D human motion tracking from monocular image," in IEEE International Conference on Acoustics, Speech and Signal Processing (Dallas, TX: IEEE), 1398–1401.

[28] Wang, H., Kläser, A., Schmid, C., and Liu, C. L. (2013). Dense trajectories and motion boundary descriptors for action recognition. Int. J. Comput. Vis. 103, 60–79. doi:10.1007/s11263-012-0594-8

[29] Yan, X., Kakadiaris, I. A., and Shah, S. K. (2014). Modeling local behaviour for predicting social interactions towards human tracking. Pattern Recognit. 47, 1626–1641. doi:10.1016/j.patcog.2013.10.019

[30] Pirsiavash, H., and Ramanan, D. (2012). "Detecting activities of daily living in firstperson camera views," in Proc. IEEE Computer Society Conference on Computer Vision and Pattern Recognition (Providence, RI), 2847–2854.

[31] Gan, C., Wang, N., Yang, Y., Yeung, D. Y., and Hauptmann, A. G. (2015). "DevNet: a deep event network for multimedia event detection and evidence recounting," in Proc. IEEE Computer Society Conference on Computer Vision and Pattern Recognition (Boston, MA), 2568–2577.

[32] Jainy, M., Gemerty, J. C., and Snoek, C. G. M. (2015). "What do 15,000 object categories tell us about classifying and localizing actions?," in Proc. IEEE Computer Society Conference on Computer Vision and Pattern Recognition (Boston, MA), 46–55.

[33] C. R. Wren, A. Azarbayejani, T. Darrell, and A. P. Pentland, "Pfinder: real-time tracking of the human body," IEEE Transactions on Pattern Analysis and Machine Intelligence, vol. 19, pp. 780–785, 1997.

[34] O. Oreifej and Z. Liu, "Hon4d: histogram of oriented 4d normals for activity recognition from depth sequences," in 2013 IEEE Conference on Computer Vision and Pattern Recognition, pp. 716–723, Portland, OR, USA, 2013.

[35] J. Shotton, T. Sharp, A. Kipman et al., "Real-time human pose recognition in parts from single depth images," Communications of the ACM, vol. 56, pp. 116–124, 2013.

[36] W. Li, Z. Zhang, and Z. Liu, "Action recognition based on a bag of 3d points," in 2010 IEEE Computer Society Conference on Computer Vision and Pattern, pp. 9–14, San Francisco, CA, USA, 2010

[37] Shabani, A. H., Clausi, D., and Zelek, J. S. (2011). "Improved spatio-temporal salient feature detection for action recognition," in Proc. British Machine Vision Conference (Dundee), 1–12.

[38] Li, R., and Zickler, T. (2012). "Discriminative virtual views for cross-view action recognition," in Proc. IEEE Computer Society Conference on Computer Vision and Pattern Recognition (Providence, RI), 2855–2862

[39] Li, B., Camps, O. I., and Sznaier, M. (2012). "Cross-view activity recognition using hankelets," in Proc. IEEE Computer Society Conference on Computer Vision and Pattern Recognition

(Providence, RI), 1362–1369
[40] Vrigkas, M., Karavasilis, V., Nikou, C., and Kakadiaris, I. A. (2013). "Action recognition by matching clustered trajectories of motion vectors," in Proc. International Conference on Computer Vision Theory and Applications (Barcelona), 112–117.
[41] Lan, T., Wang, Y., and Mori, G. (2011). "Discriminative figure-centric models for joint action localization and recognition," in Proc. IEEE International Conference on Computer Vision (Barcelona), 2003–2010.
[42] Iosifidis, A., Tefas, A., and Pitas, I. (2012a). Activity-based person identification using fuzzy representation and discriminant learning. IEEE Trans. Inform. Forensics Secur. 7, 530–542. doi:10.1109/TIFS.2011.2175921
[43] Morariu, V. I., and Davis, L. S. (2011). "Multi-agent event recognition in structured scenarios," in Proc. IEEE Computer Society Conference on Computer Vision and Pattern Recognition (Colorado Springs, CO), 3289–3296.
[44] Chen, C. Y., and Grauman, K. (2012). "Efficient activity detection with maxsubgraph search," in Proc. IEEE Computer Society Conference on Computer Vision and Pattern Recognition (Providence, RI), 1274–1281
[45] Sigal, L., Isard, M., Haussecker, H., and Black, M. J. (2012b). Loose-limbed people: estimating 3D human pose and motion using non-parametric belief propagation. Int. J. Comput. Vis. 98, 15–48. doi:10.1007/s11263-011-0493-4
[46] Tran, K. N., Kakadiaris, I. A., and Shah, S. K. (2012). Part-based motion descriptor image for human action recognition. Pattern Recognit. 45, 2562–2572. doi:10.1016/j.patcog.2011.12.028
[47] Wu, Q., Wang, Z., Deng, F., Chi, Z., and Feng, D. D. (2013). Realistic human action recognition with multimodal feature selection and fusion. IEEE Trans. Syst. Man Cybern. Syst. 43, 875–885. doi:10.1109/TSMCA.2012.2226575
[48] Liu, N., Dellandréa, E., Tellez, B., and Chen, L. (2011b). "Associating textual features with visual ones to improve affective image classification," in Proc. International Conference on Affective Computing and Intelligent Interaction, Lecture Notes in Computer Science, Vol. 6974 (Memphis, TN), 195–204.
[49] Martinez, H. P., Yannakakis, G. N., and Hallam, J. (2014). Don't classify ratings of affect; rank them! IEEE Trans. Affective Comput. 5, 314–326. doi:10.1109/ TAFFC.2014.2352268
[50] Song, Y., Morency, L. P., and Davis, R. (2012a). "Multimodal human behaviour analysis: learning correlation and interaction across modalities," in Proc. ACM International Conference on Multimodal Interaction (Santa Monica, CA), 27–30
[51] Vrigkas, M., Nikou, C., and Kakadiaris, I. A. (2014b). "Classifying behavioural attributes using conditional random fields," in Proc. 8th Hellenic Conference on Artificial Intelligence, Lecture Notes in Computer Science, Vol. 8445 (Ioannina), 95–104
[52] Patron-Perez, A., Marszalek, M., Reid, I., and Zisserman, A. (2012). Structured learning of human interactions in TV shows. IEEE Trans. Pattern Anal. Mach. Intell. 34, 2441–2453. doi:10.1109/TPAMI.2012.24
[53] Marín-Jiménez, M. J., Noz Salinas, R. M., Yeguas-Bolivar, E., and de la Blanca, N. P. (2014). Human interaction categorization by using audio-visual cues. Mach. Vis. Appl. 25, 71–84. doi:10.1007/s00138-013-0521-1
[54] Kong, Y., Jia, Y., and Fu, Y. (2014a). Interactive phrases: semantic descriptions for human interaction recognition. IEEE Trans. Pattern Anal. Mach. Intell. 36, 1775–1788. doi:10.1109/TPAMI.2014.2303090
[55] Wang, S., Ma, Z., Yang, Y., Li, X., Pang, C., and Hauptmann, A. G. (2014). Semisupervised multiple feature analysis for action recognition. IEEE Trans. Multimedia 16, 289–298. doi:10.1109/TMM.2013.2293060
[56] Matikainen, P., Hebert, M., and Sukthankar, R. (2009). "Trajectons: action recognition through the motion analysis of tracked features," in Workshop on Video-Oriented Object and Event Classification, in Conjunction with ICCV (Kyoto: IEEE), 514–521
[57] Raptis, M., Kokkinos, I., and Soatto, S. (2012). "Discovering discriminative action parts from

mid-level video representations," in Proc. IEEE Computer Society Conference on Computer Vision and Pattern Recognition (Providence, RI), 1242–1249
[58] Y. Li, W. Li, V. Mahadevan, and N. Vasconcelos, "Vlad3: encoding dynamics of deep features for action recognition," in 2016 IEEE Conference on Computer Vision and Pattern Recognition (CVPR), pp. 1951–1960, Las Vegas, NV, USA, 2016.
[59] D. L. Vail, M. M. Veloso, and J. D. Lafferty, "Conditional random fields for activity recognition," in Proceedings of the 6th International Joint Conference on Autonomous Agents and Multiagent Systems, p. 235, Honolulu, Hawaii, 2007
[60] V. Vapnik, S. E. Golowich, and A. Smola, "On discriminative vs. generative classifiers: a comparison of logistic regression and naive bayes," in Advances in Neural Information Processing Systems, vol. 9, pp. 841–848, Vancouver, British Columbia, Canada, 1996.
[61] C. Schüldt, I. Laptev, and B. Caputo, "Recognizing human actions: a local SVM approach," in Pattern Proceedings of the 17th International Conference on Pattern Recognition, 2004. ICPR 2004, pp. 32–36, Cambridge, UK, 2004
[62] H. I. Suk, B. K. Sin, and S. W. Lee, "Hand gesture recognition based on dynamic Bayesian network framework," Pattern Recognition, vol. 43, pp. 3059–3072, 2010.
[63] J. Yamato, J. Ohya, and K. Ishii, "Recognizing human action in time-sequential images using hidden markov model," in Proceedings 1992 IEEE Computer Society Conference on Computer Vision and Pattern Recognition, pp. 379–385, Champaign, IL, USA, 1992.
[64] P. Natarajan and R. Nevatia, "Online, real-time tracking and recognition of human actions," in 2008 IEEE Workshop on Motion and Video Computing, pp. 1–8, Copper Mountain, CO, USA, 2008.
[65] S. Hongeng and R. Nevatia, "Large-scale event detection using semi-hidden markov models," in Proceedings Ninth IEEE International Conference on Computer Vision, pp. 1455–1462, Nice, France, 2003.
[66] N. M. Oliver, B. Rosario, and A. P. Pentland, "A Bayesian computer vision system for modeling human interactions," IEEE Transactions on Pattern Analysis and Machine Intelligence, vol. 22, pp. 831–843, 2000
[67] T. Darrell and A. Pentland, "Space-time gestures," in Proceedings of IEEE Conference on Computer Vision and Pattern Recognition, pp. 335–340, New York, NY, USA, 1993
[68] A. Bobick and J. Davis, "An appearance-based representation of action," in Proceedings of 13th International Conference on Pattern Recognition, pp. 307–312, Vienna, Austria, 1996.
[69] A. F. Bobick and J.W. Davis,"The recognition of humanmovement using temporal templates,"IEEE Transactions on Pattern Analysis and Machine Intelligence, vol. 23, pp. 257–267, 2001
[70] Sargin, M. E., Yemez, Y., Erzin, E., and Tekalp, A. M. (2007). Audiovisual synchronization and fusion using canonical correlation analysis.IEEE Trans. Multimedia 9, 1396–1403. doi:10.1109/TMM.2007.906583
[71] Candamo, J., Shreve, M., Goldgof, D. B., Sapper, D. B., and Kasturi, R. (2010). Understanding transit scenes: a survey on human behaviour-recognition algorithms. IEEE Trans. Intell. Transp. Syst. 11, 206–224. doi:10.1109/TITS.2009. 2030963
[72] Atrey, P. K., Hossain, M. A., El-Saddik, A., and Kankanhalli, M. S. (2010). Multimodal fusion for multimedia analysis: a survey. Multimed. Syst. 16, 345–379. doi:10.1007/s00530-010-0182-0
[73] Shivappa, S., Trivedi, M. M., and Rao, B. D. (2010). Audiovisual information fusion in human-computer interfaces and intelligent environments: a survey. Proc. IEEE 98, 1692–1715. doi:10.1109/JPROC.2010.2057231
[74] Karpathy, A., Toderici, G., Shetty, S., Leung, T., Sukthankar, R., and Fei-Fei, L. (2014). "Large-scale video classification with convolutional neural networks," in Proc. IEEE Computer Society Conference on Computer Vision and Pattern Recognition (Columbus, OH), 1725–1732
[75] Westerveld, T., de Vries, A. P., van Ballegooij, A., de Jong, F., and Hiemstra, D. (2003). A probabilistic multimedia retrieval model and its evaluation. EURASIP J. Appl. Signal Process.

2003, 186–198. doi:10.1155/S111086570321101X
[76] Jiang, B., Martínez, B., Valstar, M. F., and Pantic, M. (2014). "Decision level fusion of domain specific regions for facial action recognition," in Proc. International Conference on Pattern Recognition (Stockholm), 1776–1781.
[77] Karen Simonyan and Andrew Zisserman. Two-stream convolutional networks for action recognition in videos. In Advances in neural information processing systems, pages 568– 576, 2014
[78] Mahrishi, M., Hiran, K. K., Meena, G., & Sharma, P. (2020). Machine Learning and Deep Learning in Real-Time Applications. IGI Global
[79] Efros et al., 2003; Schuldt et al., 2004; Jhuang et al., 2007; Fathi and Mori, 2008; Niebles et al., 2008
[80] Christoph Feichtenhofer, Axel Pinz, and Andrew Zisserman. Convolutional two-stream network fusion for video action recognition. In Proceedings of the IEEE Conference on Computer Vision and Pattern Recognition, pages 1933–1941, 2016
[81] Chuang Gan, Chen Sun, Lixin Duan, and Boqing Gong. Webly-supervised video recognition by mutually voting for relevant web images and web video frames. In European Conference on Computer Vision, pages 849–866. Springer, 2016
[82] Chuang Gan, Ting Yao, Kuiyuan Yang, Yi Yang, and Tao Mei. You lead, we exceed: Labor-free video concept learning by jointly exploiting web videos and images. In Proceedings of the IEEE Conference on Computer Vision and Pattern Recognition, pages 923–932, 2016
[83] Rohit Girdhar, Deva Ramanan, Abhinav Gupta, Josef Sivic, and Bryan Russell. Actionvlad: Learning spatio-temporal aggregation for action classification. In CVPR, volume 2, page 3, 2017
[84] Andrej Karpathy, George Toderici, Sanketh Shetty, Thomas Leung, Rahul Sukthankar, and Li Fei-Fei. Large-scale video classification with convolutional neural networks. In Proceedings of the IEEE conference on Computer Vision and Pattern Recognition, pages 1725–1732, 2014
[85] Nitish Srivastava, Elman Mansimov, and Ruslan Salakhudinov. Unsupervised learning of video representations using lstms. In International conference on machine learning, pages 843–852, 2015
[86] Du Tran, Lubomir Bourdev, Rob Fergus, Lorenzo Torresani, and Manohar Paluri. Learning spatiotemporal features with 3d convolutional networks. In Proceedings of the IEEE international conference on computer vision, pages 4489–4497, 2015
[87] Limin Wang, Yuanjun Xiong, Zhe Wang, Yu Qiao, Dahua Lin, Xiaoou Tang, and Luc Van Gool. Temporal segment networks: Towards good practices for deep action recognition. In European Conference on Computer Vision, pages 20–36. Springer, 2016
[88] Xiaolong Wang, Ross Girshick, Abhinav Gupta, and Kaiming He. Non-local neural networks. arXiv preprint arXiv:1711.07971, 10, 2017
[89] Saining Xie, Chen Sun, Jonathan Huang, Zhuowen Tu, and Kevin Murphy. Rethinking spatiotemporal feature learning: 7092 Speed-accuracy trade-offs in video classification. In Proceedings of the European Conference on Computer Vision (ECCV), pages 305–321, 2018
[90] Joe Yue-Hei Ng, Matthew Hausknecht, Sudheendra Vijayanarasimhan, Oriol Vinyals, Rajat Monga, and George Toderici. Beyond short snippets: Deep networks for video classification. In Proceedings of the IEEE conference on computer vision and pattern recognition, pages 4694–4702, 2015
[91] Christopher Zach, Thomas Pock, and Horst Bischof. A duality based approach for realtime tv-l 1 optical flow. In Joint Pattern Recognition Symposium, pages 214–223. Springer, 2007
[92] Bolei Zhou, Alex Andonian, and Antonio Torralba. Temporal relational reasoning in videos. arXiv preprint arXiv:1711.08496, 2017
[93] Mohammadreza Zolfaghari, Kamaljeet Singh, and Thomas Brox. Eco: Efficient convolutional network for online video understanding. arXiv preprint arXiv:1804.09066, 2018
[94] Feng, S.; and Duarte, M. F. 2019. Few-Shot LearningBased Human Activity Recognition. arXiv preprint arXiv:1903.10416
[95] Shin, H.-C.; Roth, H. R.; Gao, M.; Lu, L.; Xu, Z.; Nogues, I.; Yao, J.; Mollura, D.; and Summers,

R. M. 2016. Deep convolutional neural networks for computer-aided detection: CNN architectures, dataset characteristics and transfer learning. IEEE transactions on medical imaging 35(5): 1285–1298

[96] Salem, M.; Taheri, S.; and Yuan, J.-S. 2018. ECG Arrhythmia Classification Using Transfer Learning from 2-Dimensional Deep CNN Features. In 2018 IEEE Biomedical Circuits and Systems Conference (BioCAS), 1–4. IEEE

[97] Salem, M.; Taheri, S.; and Yuan, J.-S. 2018. ECG Arrhythmia Classification Using Transfer Learning from 2-Dimensional Deep CNN Features. In 2018 IEEE Biomedical Circuits and Systems Conference (BioCAS), 1–4. IEEE

[98] Mahrishi M., Morwal S. (2020) "Index Point Detection and Semantic Indexing of Videos—A Comparative Review". In: Pant M., Kumar Sharma T., Arya R., Sahana B., Zolfagharinia H. (eds) Soft Computing: Theories and Applications. Advances in Intelligent Systems and Computing, vol 1154. Springer, Singapore. https://doi.org/10.1007/978-981-15-4032-5_94

[99] Heldman, D. A.; et al. 2017. Telehealth Management of Parkinson's Disease Using Wearable Sensors: an Exploratory Study. Digital biomarkers 1(1): 43–51

[100] Morcos, A.; Raghu, M.; and Bengio, S. 2018. Insights on representational similarity in neural networks with canonical correlation. In Advances in Neural Information Processing Systems, 5732–5741

[101] Devlin, J., Chang, M.W., Lee, K., Toutanova, K.: BERT: Pre-training of Deep Bidirectional Transformers for Language Understanding (10 2018)

[102] Feichtenhofer, C., Fan, H., Malik, J., He, K.: Slowfast networks for video recognition. In: Proceedings of the IEEE International Conference on Computer Vision. vol. 2019-October, pp. 6201–6210. Institute of Electrical and Electronics Engineers Inc. (10 2019)

IOP Conf. Series: Materials Science and Engineering 1131 (2021) 012005 doi:10.1088/1757-899X/1131/1/012005

Stock trend prediction and analysis using LSTM neural network and dual moving average crossover algorithm

Rahul Pradhan[1], Kanishk Bharadwaj[2], Aman Saxena[3] and Vikram Rajpoot[4]

GLA University, 17km Stone, NH-2, Mathura-Delhi Road Mathura, Chaumuhan, Uttar Pradesh 281406, India

[1] rahul.pradhan@gla.ac.in, [2] kanishk.bharadwaj_cs18@gla.ac.in,
[3] aman.saxena_cs18@gla.ac.in, [4] vikram.rajpoot@gla.ac.in

Abstract. In today's world where data is the new currency, computer machines are capable enough to find and predict the future by using the previous data and the algorithms which are becoming more precise day by day. The proposed solution by us includes the pre-processing of the 10 years of Bombay Stock Exchange (BSE) stock market data of HDFC Bank. Our solution implemented the engineering techniques of deep learning and machine learning which are combined with the LSTM neural network algorithm and Dual Moving Average Crossover algorithm.

1. Introduction

Investing in the stock market has always been a dream of common men who want to invest money but as we all are familiar with the fact that banks usually provide an interest rate of 3% to 4% on savings accounts which is very low but still people use this as their primary choice.

Now here the question is, why they choose banks over other choices of investments. This has a very common and traditional answer, they have the security of money in banks. This means they will get their total invested money whenever they want and with some percent of simple interest added to the total amount as well.

During investing in the share market people are usually afraid of losing money as they have almost zero knowledge about the share market and the one who is having knowledge about it is also somewhere afraid that either he would gain or totally lose money.

After analyzing the stock market, we have found that there somewhere exist some trends in the stock price of the companies in the share market. We found that there are 2 major stock markets in India where people are interested in investing which are Bombay Stock Exchange (BSE) and National Stock Exchange (NSE), NSE has 1328 listed companies out of which the top 50 companies stock price decide the performing index (NIFTY) whereas BSE has 5000 listed companies and here only the top 30 companies decide the percentage (SENSEX analysis). Hence it is clear that SENSEX is a better performance index than NIFTY.

Technical analysts and brokers generally believe that the prices proceed in trends and it tends to repeat themselves after a period when it comes to the market's overall mindset. Using the previous year's data of companies, we aim to use the past performance to predict the future trends in the stock

Content from this work may be used under the terms of the Creative Commons Attribution 3.0 licence. Any further distribution of this work must maintain attribution to the author(s) and the title of the work, journal citation and DOI.
Published under licence by IOP Publishing Ltd

market [1]. We have used HDFC Bank for this analysis which is a part of BSE and SENSEX. We have taken around 10 years of previous data (2011-2020) of HDFC Bank and performed the analysis on them. The algorithms we have used are DMAC [2] and LSTM[3] to predict the behavior of the market.

We chose the LSTM[3] neural network over RNN[4] because RNN has a drawback of looking at the recent data to perform the current task. It can only predict based on recent information and it also has a problem called long term dependency problem. That's why we used the LSTM neural network, it is designed to avoid long-term dependency problems. It can remember the information for a long period of time, basically, it's the default behavior of LSTM[5,6]. So, it can predict easily based on the previous 10 years of the dataset.

2. Related Work

Table 1. Literature Survey.

Author	Approach	Dataset Used	Pros	Cons
Kim and Han[2]	ANN and Genetic Algorithm (GA)	They used the Korean Stock Price Index (KSPI) The dataset contains 2928 days of trading data from Jan 1989 to Dec 1998	They use GA to optimize the model of ANN They use feature discretization to optimize or reduce the dimension of input data	Issue with ANN models is, it needs lots of data because of the large number of connections in ANN. Another issue with this model is that they only focus on two features of optimization
Jingyi Shen and M. Omair Shafq[1]	PCA algorithm to reduce the dimensions of input data and LSTM model to predict the stock price trend	3558 stocks from Chinese Stock Market	Research Paper uses LR algorithms which achieve the same result as SVM and MLP but take less training time as compared to other algorithms. Their LSTM model achieves 93.25% accuracy.	Research Paper only focus on short term trend prediction Research Paper does not show when to sell and buy the stocks They just predict the future price trend of Chinese stock market
Anton H.Hofman and Gerrit ten Brinke[3]	They used GA optimized and extracted features with a fuzzy	Stock Market dataset of 200 largest companies listed in the	They proposed an optimized model for selection and scoring of stocks	They did not compare the model with the pre-existing

	membership function with a model, for optimizing stock scoring.	TWSE and TEJ.	apart and distinguished with the traditional prediction model. They majorly focused on rankings, selection, and performance evaluation which are the factors one should keep in mind.	algorithms we use which made it more challenging to identify if GA would perform better than the other pre-existing algorithms.
Manoj Thakur and Deepak Kumar[7]	They used Random Forest (RM) for pruning of features in the approach. The algorithms were with a WMGESVM to generate the "Buy/Hold/Sell" signals for the analysis.	NASDAQ, NIFTY 50, S&P 500, DOW JONES and NIFTY BANK	They compared their algorithm with the pre-existing Random Forest algorithm. The used RF for feature purring which is a good thing.	The issue with their work was that they didn't mention any kind of time or space complexity of their work which showed that irrespective of the accuracy of their algorithm the time and space factor is somehow lost.
McNally S, Roche J, Caton S[8]	They used RNN and LSTM models and Boruta Algorithm which works the same as a random forest classifier	They used the BItcoin trading dataset. The dataset contains data from 19 Aug 2013 to 19 July 2016	They used various optimization to improve deep learning algorithm. Prediction of Bitcoin price same as stock market price prediction	They majorly treated questions as a time sequence problem. Their dataset contains various noise which directly affects the prediction

Cheng-LungHuang and Cheng-Yi Tsai[6]	They used filter-based feature selection which is assembled with a hybrid Self-Organizing Feature Map (SOFM)[5]	Taiwan index futures (FITX) for prediction	They also applied feature selection before the processing for improving the prediction accuracy and to reduce the computational complexity of processing the stock market data.	The algorithm could only predict for next day index price which is not feasible for long-time market investors. Moreover, the evaluation was not strong enough as they set a single SVR model as the baseline, and they didn't compare previous works with their model's performance.
Fischer T, Krauss c[7]	They use Artificial Recurrent Neural Network architecture LSTM	They used the S & P 500 Index dataset. The dataset contains data from Dec 1989 to Sept 2015	They used the latest artificial recurrent neural network architecture. Their LSTM model performs far better than Standard Deep Neural Network(DNN) and regression algorithms.	They too much relied on LSTM. They didn't mention how they trained LSTM model with long time prediction They didn't show when to buy and sell the stock. They just predicted the future price of the stock market.
Thakkar and Chaudhari[9]	By using cross-reference they predicted the price of one market with the trained data of another share market	They used BSE and NSE data.	They used the cross-reference technique which is not often used and it is a good choice	They did not use the data of companies other than IT companies. All types of companies show different trends.
M. Ananthi & K. Vijayakumar[10]	They used candlestick regression and market trend prediction (CKRM)	They used tech companies datasets	They did a good comparison among many algorithms.	They did not focus on trends analysis and they also did not take the dataset of companies other than tech companies.

3. Proposed Approach

In this work we took the last 10 years of the data of HDFC Bank from www.yahoo.com [11] and then we cleaned the dataset. We then extracted the weighted features which showed correlation. We then reshaped the data according to our needs and according to the need of the LSTM neural network.

We fed that data to the LSTM model and tested our model on that dataset followed by training our model with another splitted part of the dataset.

We then used DMAC algorithm to generate buy and sell signals and visualized those signals.

This diagram shows our proposed approach for this research work. Here we have made a flowchart to illustrate the approach.

4. Experimental Setup and Result Analysis

This section explains the ML models and algorithms used in this prediction.

We used datasets of HDFC Bank. The data used was from 01-01-2012 to 17-12-2019 dates. The dataset was imported by using .data_reader() which is a method of pandas library in Python. Furthermore we analyzed the correlation between the attributes of the dataset and reached the conclusion that taking the 'close' attribute is overall a good choice to analyze and predict the stock market trends (figure 2).

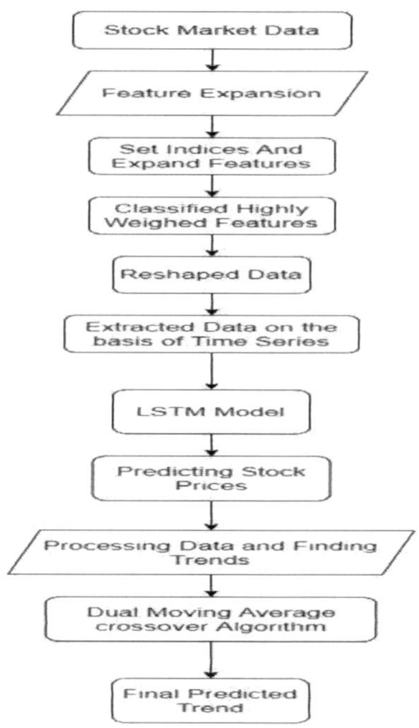

Figure 1: It shows the proposed approach of our research work

We reshaped the dataset three-dimensionally and then we transformed those values into 0 and 1 scale followed by splitting the data into 2 parts, train data, and test data. Furthermore, we split the train and test data into 2 parts, x and y train data & x and y test data.

We built the LSTM model consisting of 50 neural combined with 25 dense layers. Afterward, we trained the data by using 80% of the total data in the dataset, for all this we used the timestamp of 60 days to predict the further values.

Explanation of **Dual Moving Average Crossover Algorithm**[12] - Dual moving crossover is one of the most used trading strategies. It is exactly like the sliding windows technique which is widely used in computer programming. In this, we fix the size and slide it over the whole list, and we keep taking averages (mean value) of all the values coming inside the window.

Suppose we have a list [1,5,7,8,2,5,8,5,1,5,7,9,2,5] and size is 6, then the average will be like. (1+5+7+8+2+5+8)/6 and then the values will be (5+7+8+2+5+8+5)/6 ……. up to the size of the list.

DMAC algorithm is used by 2 windows one of size 30 and another of size 100.

Simple Moving Average (SMA) is a technique of DMAC. Whenever the SMA30 becomes greater than SMA100, then it is a signal to buy the stocks and if it becomes less than SMA100, then it signals to sell the stocks.

Figure 2: Date vs. Closing Price in INR

5. Result

The trend was analyzed by using the predicted 30 days' values and previous values of the stock market.

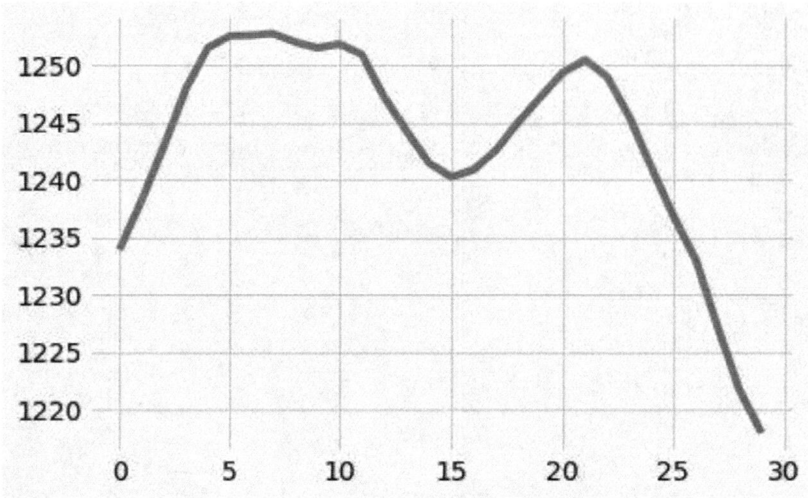

Figure 3: Days of Months vs. Stock Price

This picture figure 3 shows the days (x-axis) vs stock price (y-axis) of predicted 30 days price.

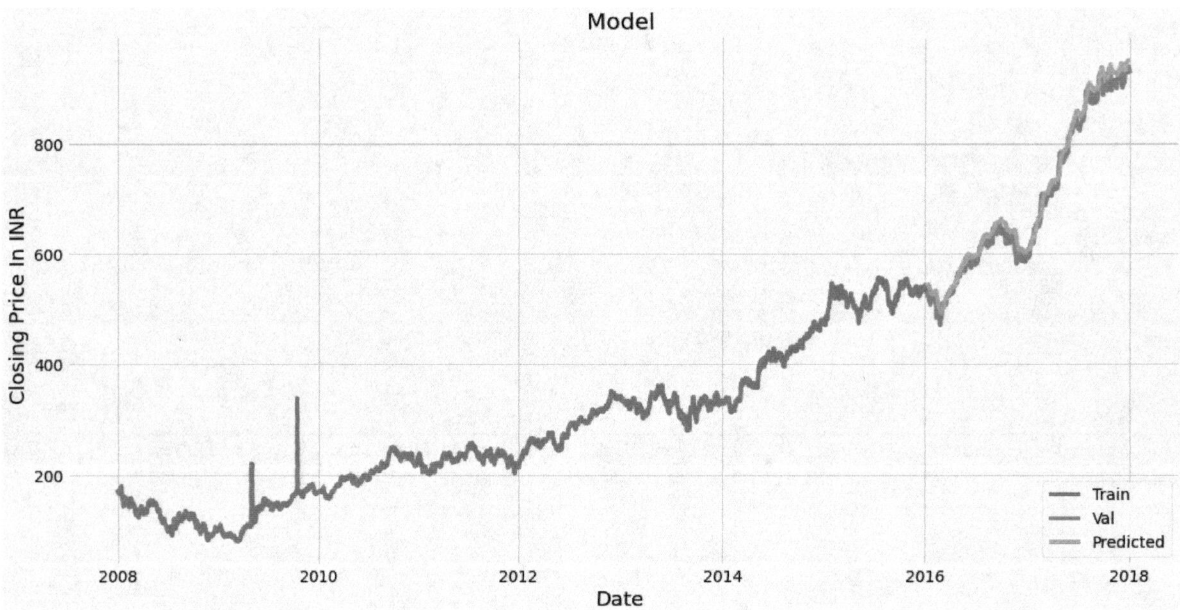

Figure 4: Years vs. Closing Price in INR (with train, Val, and Predicted Data)

The above picture figure 4 shows the date (x-axis) vs closing value (y-axis) where the blue color line shows the train values, orange color lines show the actual values and yellow color shows the predicted values. Here we found the accuracy of our LSTM model to be 94.8%.

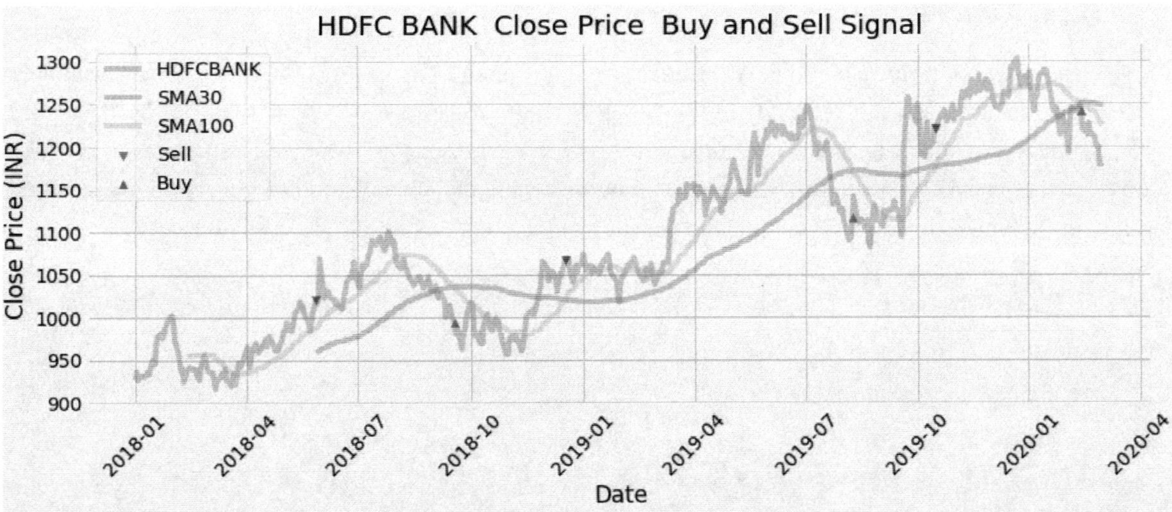

Figure 5: Years vs. Close Price (DMAC algorithm graph wrt. Historical and Predicted Price

The above graph figure 5 is a DMAC graph which shows the values of SMA30 and SMA100 with the values of HDFC Bank's original values of stocks and HDFC Bank's predicted values of the stock price.

The above graph (figure 5) shows the legends to buy and sell the stocks by the use of the Dual Moving Average Crossover (DMAC) algorithm.

In this graph (figure 5), the red color down arrow shows when to sell the stocks and the green color up arrow shows when to buy the stocks. These both generated signals are at the intersection points of SMA30 and SMA100. The detailed conclusion from this graph is in the conclusion section.

5.1 Result Table

Table 2. Evaluation Results.

Topic	Detailed Explanation	Remarks and Results
LSTM neural Network	We used this to perform the prediction for stock market prices. We first predicted the values for the years' data we already had and then we found the accuracy of the model to be 94.8%. We then used that model to predict the prices of the months of which we were not having data.	We found that accuracy of the model is 94.8% which shows that the model is very efficient.
DMAC Algorithm	We used this algorithm to generate signals. For further detailed explanation of the DMAC algorithm refer to [9].	We found that this algorithm is very efficient and there was not any point in the graph where we can see that generated signals are wrong.

6. Conclusion

Our work was based on predicting the stock market price with the use of the DMAC algorithm and LSTM neural network. We achieved an accuracy of 94.8% from which we can conclude that analyzing trends in the Stock Market is also a good choice and a lot of research work can be done in this aspect of the stock market.

6.1 Justification of Results

By visualizing the figure 5 we found that Dual Moving Average Crossover (DMAC) is also an accurate and efficient way to analyze trends in the Stock Market. We also found that investing in the stock market using trends analysis (or we can say long trends) is a very low risk task or an almost zero risk task as the signals are very accurate and we couldn't find any case where the generated signals are showing wrong indications of buying and selling, which could result into the loss of money.

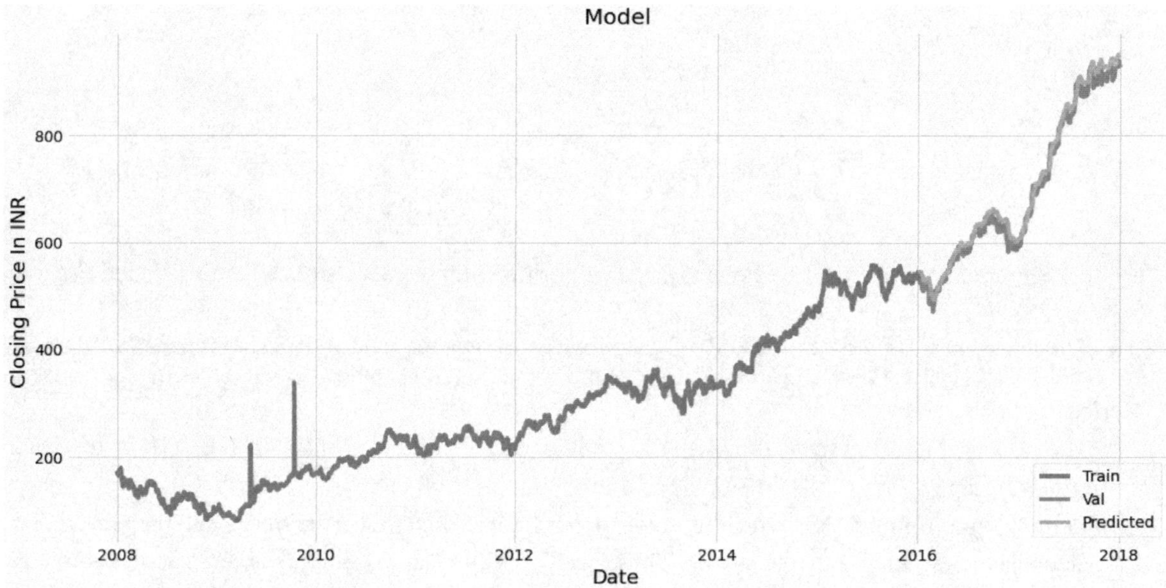

Figure 6: Years vs. Closing Price in INR (with train, Val, and Predicted Data)

It is very clear from this graph (figure 6) that the predicted values and original values are very near to each other which shows that trends also play a major role in predicting the prices of the stock market just exactly like sentiments do.

7. **Future Scope**

The future work which we will do in this research work is, we will compare other machine learning algorithms like Random forest classifier and multiple regression etc. and other trend analysis algorithms like EMA, MACD and WMA etc. to find the differences between them and also we will find which is the most efficient algorithm for different markets.

We will also use datasets of other stock markets and datasets of different types of companies to find if there exists no pattern or some unique patterns in those companies and markets.

References

[1] Shen J and Shafiq M O 2020 Short-term stock market price trend prediction using a comprehensive deep learning system J. Big Data 7

[2] Kim K-J and Han I 2000 Genetic algorithms approach to feature discretization in artificial neural networks for the prediction of stock price index Expert Syst. Appl. 19 125–32

[3] Olah, C., 2021. *Understanding LSTM Networks -- colah's blog*. [online] Colah.github.io. Available at: <http://colah.github.io/posts/2015-08-Understanding-LSTMs/> [Accessed 10

February 2021].
[4] Hofman A H, ten Brinke G and Loos K 2016 Hierarchical structure formation in supramolecular comb-shaped block copolymers Polymer (Guildf.) 107 343–56
[5] Mahrishi, M., Hiran, K. K., Meena, G., & Sharma, P. (2020). Machine Learning and Deep Learning in Real-Time Applications. IGI Global.
[6] Huang C-L and Tsai C-Y 2009 A hybrid SOFM-SVR with a filter-based feature selection for stock market forecasting Expert Syst. Appl. 36 1529–39
[7] Fischer T and Krauss C 2018 Deep learning with long short-term memory networks for financial market predictions Eur. J. Oper. Res. 270 654–69
[8] Thakur M and Kumar D 2018 A hybrid financial trading support system using multi-category classifiers and random forest Appl. Soft Comput. 67 337–49
[9] Thakkar A and Chaudhari K 2020 Predicting stock trend using an integrated term frequency–inverse document frequency-based feature weight matrix with neural networks Appl. Soft Comput. 96 106684.
[10] Ananthi M and Vijayakumar K 2020 Stock market analysis using candlestick regression and market trend prediction (CKRM) J. Ambient Intell. Humaniz. Comput.
[11] McNally S, Roche J and Caton S 2018 Predicting the price of bitcoin using machine learning 2018 26th Euromicro International Conference on Parallel, Distributed and Network-based Processing (PDP) (IEEE) pp 339–43
[12] Kumar G, Jain S and Singh U P 2020 Stock market forecasting using computational intelligence: A survey Arch. Comput. Methods Eng.

Minimizing Respiratory Diseases using Warning from Hand-Touch-Over-Face Avoidance Based Algorithm on Deep Learning

Jitesh Kumar Bhatia

Department of Computer Engineering and Applications,
GLA University, Mathura

jitesh.bhatiya@gla.ac.in

Abstract. The most moved parts of a human body are hands. They may get contaminated by touching any surface we are working with. The contaminated hands can intentionally or unintentionally reach the part of our face while eating, sneezing, setting of hair, itching a scratch, biting the nails, soothing down when stressed, etc. The touch of hands in eyes, nose and mouth may result in unhygienic symptoms and may result in various respiratory diseases. We need to minimize our touch to the parts of our face. In the presented work, we propose a system that warns the user each time his/her hand reaches his/her face. We have used CNN to identify the class feature automatically rather than manually. Also, we have used RNN so that the proposed system learns recursively. The use of such system may minimize the subsequent touch on face and will lead to minimizing the respiratory diseases due to unhygienic touch conditions.
Keywords- Hand Over Face; Respiratory Diseases; Convolutional Neural Networks; Recurrent Neural Networks; Machine Learning.

1. Introduction

The hands are the most moved organ of a human body. Also, hands are used by all the human beings to feel the touch of any surface. It results in the transfer of Virus and bacteria from a surface to our hands. This usually creates unhygienic symptoms in hands. These hands may reach our face while eating, sneezing, touching and/or setting the hair, itching in eyes, wiping tears, cleaning the mucous, to scratch an itch, to soothe down when stressed, biting nails, etc. to name a few. The contaminated hands, when touch the mouth, nose or eyes can result in infections such as influenza, coronavirus, Staphylococcus aureus (a disease carried in nasal nucosa), conjunctivitis, etc. In a nutshell, the infection in a human being is the result of self-inoculation. Self-inoculation is a type of contact transmission where a person's contaminated hands makes subsequent contact with other body sites on oneself and introduces contaminated material to those sites [1][2]. According to a study by [3], on closely monitoring 26 students, for duration of 240 minutes, they, in an hour on an average touched their face 23 times. The average duration of mouth touching was 2 seconds whereas the average nose touching duration was 1 second. Figure 1 shows the average touch of hands on each significant part of the face. It was even notified that the urge to touch the face was increased if the movement of hands was prevented knowingly.

Many solutions to avoid the contact of hands with the face have already been given by different researchers. Some of them are identification the movement of hands with the help of sensor-based-gloves, others being identified using the wearables such as a watch or a ring using IoT, whereas some of them identified through an algorithm applied to the live video through a device having camera fitted

Content from this work may be used under the terms of the Creative Commons Attribution 3.0 licence. Any further distribution of this work must maintain attribution to the author(s) and the title of the work, journal citation and DOI.
Published under licence by IOP Publishing Ltd

to it. Under the Computer Vision umbrella, hand segmentation is an emerging topic to be explored due to the various challenges like skin colour, light variations, complex backgrounds, etc. to name a few.

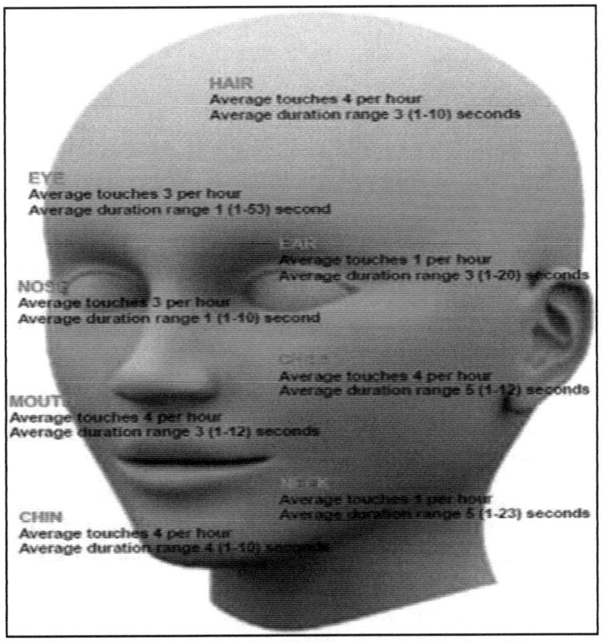

Figure 1: Average touch of hand on parts of a face [3]

In the presented work, the method deals with the identification of a hand or a part of hand (finger, wrist, etc.) on the parts of the face knowingly or unknowingly which may cause a reason for contaminated eyes, nose or mouth. The motivation behind the work is the increase in the spread of respiratory diseases. One of the major reasons of the increase is the touch of contaminated hands on the major sense organs i.e. mouth, nose and eyes. We apply the Deep Learning algorithms to warn the user in the live video as and when he/she puts his/her hand over the face. This is done by producing a beep sound and highlighting the localization of the hand on the face. In this work, we first determine each pixel of the image/frame under consideration to identify whether it is a part of a hand or not. Also, by the use of Canny-Edge Detection [4] and Viola Jones Face Detection [5], we identify the edges of the face of the user. Then, if the localization of the hand is overlapping the area of the face, we may conclude that the hand is overlapping the face. However, if there is a distance in the y-z plane or the x-z plane between hand and the face, the result is negative.

2. Related work

Much of the work has been done in the identification of face in a video and segmentation of hands in image and video. However, hand over face segmentation is an area of Computer vision that still requires a plenty of efforts. To an extent, there are basically two approaches that can be followed to detect hand over a face in a 3-dimensional environment. Firstly, having sensors fitted in the hand in the form of wearable, that detects the localization of the wearable (sensors) on the face. Secondly, by having a vision based technique that identifies the presence of a hand over the face. The first technique being costly due to the involvement of hardware like sensors and hardware required for implementing IoT, the latter technique is favored by most of the researchers.

The process of determining the class/category of each pixel of an image is known as semantic segmentation. The semantic segmentation basically involves either of the two methods, namely, the probabilistic approach and the deep learning technique. Though the traditional approach was the probabilistic approach, recently, the latter one is outperforming better for semantic segmentation.

Out of the many public-datasets available on the internet, the most popular ones related to the hands-over-face are VLM-HandOverFace dataset [6] and the HOF dataset [7]. The VLM-

HandOverFace dataset [6] contains more than 4000 images and is available for research for academicians. The HOF dataset [7] has been created using the random images on the internet. The dataset contains images with hands near and in front of the face. The size of the data set is 300 frames. It consists of images of people with different ethnic backgrounds, color, age and gender.

Some of the related works are described as under.

2.1. Sensor based Hand over Face Segmentation

Sudarshan B, et. al [8] designed a model for smart watches to identify the hand movement to the face by training the classifier using 2071 patterns. Out of these, 102 features were extracted to describe the movement of hands towards the face. They applied one-class SVM, iForest, Minimum Covariance Determinant and Local-Outlier Factor on the feature extracted above. For the purpose of training the classifier, they used the combination of Accelerometer data, Gyroscope data, Pressure data and Rotation Vector data. They ignored the non-hand to face movements (outliers) and achieved an F1 score of 0.93.

2.2. Computer Vision based Hand over Face Segmentation

Mahmoud M, et. al [9] detected the hand when it occluded a face. They divided each frame into 9 regions keeping face in the center, and applied LBPs or Gabor Filter for the extraction of the required feature vector. Each frame was then compared with the initial frame which included only the face. In case of any discrepancy due to a regional change in the structure of the image, they generated a probability that the change was due to the hand occluding the face. Gonzalez et al. [10] tracked the segmentation of hand over face for the identification of sign language. The authors used color and edge information for their proposed methodology. Grafsgaard et al [11] identified human gestures with the presence of two hands in the image under consideration using depth images. The authors tested their research in a computer-meditated tutoring environment. Mahmoud M et al. [12] classified various gesture descriptors such as hand shape, hand action and the region of face occluded by hand. The authors used HOG, PCA and SVM to improve the performance of the proposed approach. They concluded through their results that the multi-modal linear-SVM classifier outperformed the uni-modal linear SVM classifier. Naik N, et. al [13] used deep learning approach for the purpose of identifying gestures of face when face was occluded by hand and determined the type of emotion(s) of the human being under consideration in the image. The authors used RNN to recursively learn the features to identify a broader number of emotions categories other than being happy, sad, disgust, thinking, boredom, surprise, etc. to name a few.

3. Proposed Hand-Touch-Over-Face-Avoidance Based Method

As per the discussion in the above literature, we have seen a significant amount of work has been done in the field of Hand-Over-Face-Segmentation. However, the touch of the hand on face has not been addressed by most of the researchers. In this section, we introduce our new approach for the identification of hands over face with the hands touching the face or a part of face, which is to be avoided. We not only segment the hand over face in the x-y plane but also check that the distance between the hands and face in the x-z plane or in the y-z plane is zero or minimum. Moreover, the proposed method uses the Deep Learning Technique for more accurate feature extraction and classification of the parts of the body.

The Figure 2 illustrates the basic flow of the proposed algorithm to find the hand over part of a face with less than allowed minimum distance in the 3-dimensional plane.

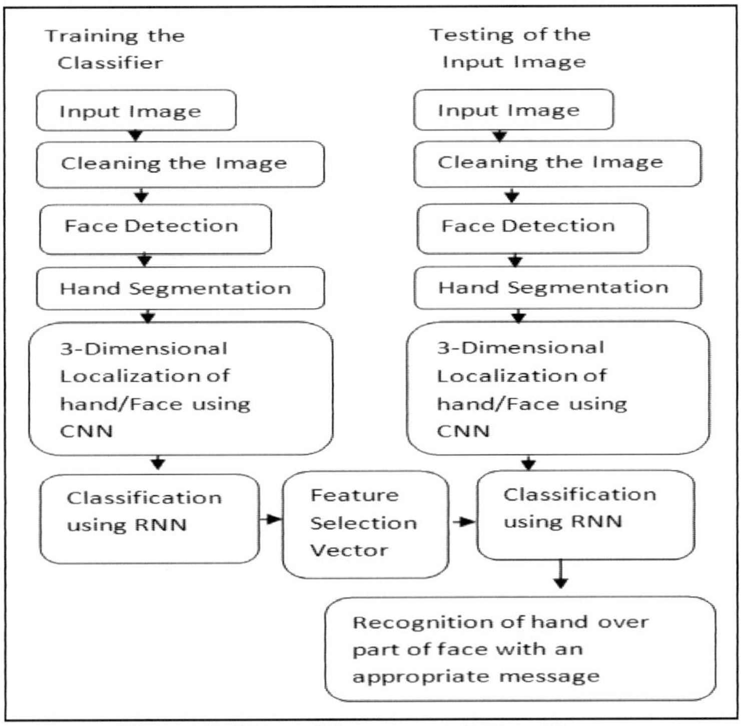

Figure 2: Flow of the proposed Algorithm

```
Input: Dataset I_train = {I1, I2, I3,...., In} of n images and image I
Output: Beep if Hand is touching the part of Face
BEGIN
//Phase1: Training the Classifier//
1. for i=1 to n images
2.          gi = rgb2gray (Ii)
3.          face_detection = Detect_Face (gi)
4.          if face_detection= 1
                      hand_detection=Detect_Hand(gi)
                      if hand_detection=1
                                 Apply CNN to extract advanced related features of Hand and Face simultaneously
                                 if 3-dimensional localization of hand and face coincide
                                            Apply RNN to add a category of warning
5. else
            No human face detected
//Phase2: Image to be Tested//
6. g = rgb2gray (I)
7. flag = Detect_Face (g)
8. if face_detection= 1
                      hand_detection=Detect_Hand(g)
                      if hand_detection=1
                                 Apply CNN to extract advanced related features of Hand and Face simultaneously
                                 if 3-dimensional localization of hand and face coincide
                                            Apply RNN to recognize category of warning from the classified categories
                                            Play a beep sound of warning
9. else
            No human face detected
END
```

Figure 3: Pseudo Code of the proposed Algorithm

The Figure 3 shows the pseudo-code for the identification of a hand over part of the face with less than allowed minimum distance in the 3-dimensional plane between them. It also produces a warning sound with an appropriate message as soon as the detection is successful.

Table 1: Notations and their description

Notation	Description of the used Notation
N	Number of Images in the Training Set
I_{train}	Set of Images used to train the classifier
I	The Image under consideration
Detect_Face	Algorithm to Detect Face
Detect_Hand	Algorithm to Detect Hand

The proposed algorithm uses certain notations which are explained in Table 1. As shown in the figure, the algorithm consists of 2-phases, namely Training and Testing. In the first phase, the classifier is being trained using n number of images in the training data-set. Each time a hand over the face is found, it is stored in the set of categories with appropriate message using RNN. After the training phase is over, the testing is performed on image I. If the image I contains a human face, it checks for hands and the localization of hands. If the localization of both, face and hand coincide, the appropriate message is shown with a beep (warning sound) that was categorized using RNN in the training phase.

The steps in each phase are described as under:

3.1. Face Detection

The beginning of the proposed algorithm is with the detection of a human face in the image. The detection of a face in an image is somewhat cumbersome due to the difference in the parameters in the image such as size and color of the face, the illumination when the image was taken, the texture that was in the vicinity, the shape of the face (oval, square-shaped, etc.) and so on. So, it is very important to choose one of the best Face Detection algorithms from the literature. As per our requirements, we have chosen Viola-Jones face detector [5] due to the advantages as stated: It is faster in determining the face and can process up to 15 frames per second and uses the integrated learning algorithm known as ADABOOST to select the appropriate face features from the ones extracted from the face.

3.2. Hand Detection

Detecting Hands in an Image is one of the major challenging tasks as the shape of the hand can be changed as per the user. The detection of hands can be affected by the difference in lightning conditions in the image and the noise in the background. The skin color also adds a challenge to the detection and extraction of a hand in an image. In our method, we have used the Canny Edge Detection to identify the hand area and the localization of the hand in the image. Also, the automatic feature extraction technique known as CNN (Convolutional Neural Network) [14][15] has been used to extract the necessary features and discard the other features. CNN mainly consists of five layers namely the input layer, Convolutional Layer, Rectified Linear Unit, Pooling Layer and the Softmax layer as shown in Figure 4. Each layer has its own function of extraction of features from the image under consideration.

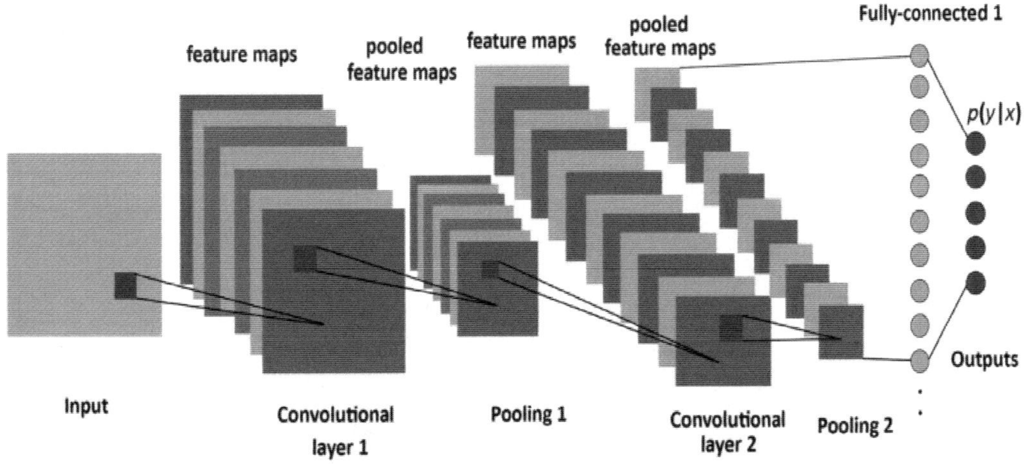

Figure 4: Architecture of CNN [15]

3.3. Classification

The third step of the algorithm i.e. classification has been performed using RNN. Recurrent Neural Network has the tendency to categorize more categories of hand over parts of a face and gives appropriate result each time a hand is found on a part of a face. The RNN also learns the features from the existing features recursively. Thus, as more and more new images are tested using RNN, the accuracy of the algorithm goes on improving.

4. Results

We have performed our algorithm on two chosen datasets namely VLM-HandOverFace dataset and the HOF dataset. The former consists of 4300 images of faces with and without hands and the latter contains 300 images randomly selected from the Internet.

True-Positives and False-Positives are basically used to detect the number of images where the warning was issued. True positives are the number of images with hands on the face successfully identified the same. However, the number of images with face only, identified occluded with hand is termed as False positives. Some of the images with hands on the face were missed to be identified by the classifier. Such images in the dataset are termed as False Negatives. After calculation of them, the Precision, Recall and F1-score can be calculated for comparison with the other approaches. The parameters are calculated as follows:

$$Precision, p_r = \frac{t_p}{t_p + f_p} \quad (1)$$

$$Recall, r_c = \frac{t_p}{t_p + f_n} \quad (2)$$

$$F_1 = \frac{2p_r r_c}{p_r + r_c} \quad (3)$$

On application on the proposed algorithm in the datasets, the values calculated as per equations (1), (2) and (3) are shown in Table 2 and Table 3 respectively.

Table 2. Experiment Results by the proposed approach on [6].

Parameters	Value

Total Number of Images in the dataset	4300
True Positive, t_p	3256
False Positive, f_p	409
True Negatives, t_n	374
Precision, p_r	.8884
Recall, r_c	.8969
F1-measure	.8926

Table 3. Experiment Results by the proposed approach on [7].

Parameters	Value
Total Number of Images in the dataset	300
True Positive, t_p	231
False Positive, f_p	34
True Negatives, t_n	27
Precision, p_r	.8716
Recall, r_c	.8953
F1-measure	.8833

The data can also be depicted in the graphical representation in the form of a bar graph shown in Figure 5.

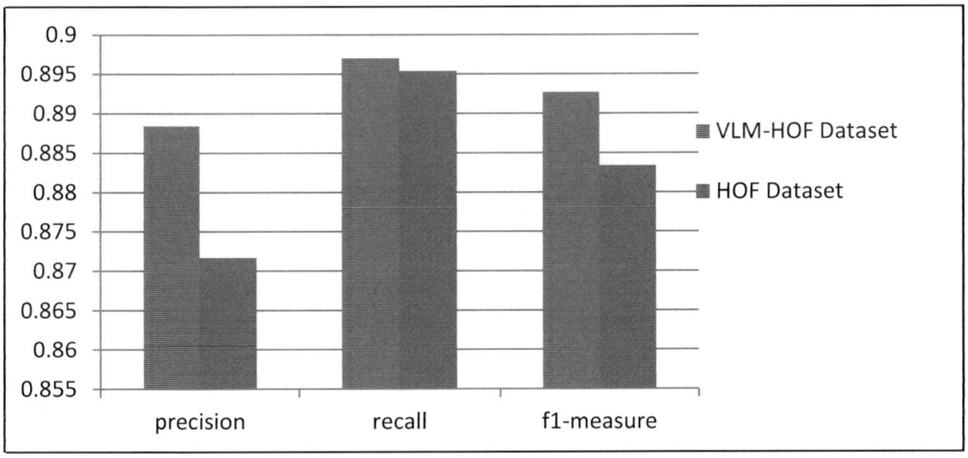

Figure 5: Graphical Representation of the Results

The above results depicts that the proposed method outperforms in the case of both the chosen data-sets with the F1-score of .89 and .88 respectively.

5. Conclusion

In the presented paper, we have proposed an algorithm which can continuously keep a check on a person who is in front of a camera such as web-cam. The person is getting continuously monitored and as soon as he/she puts his/her hand on the part of his/her face (intentionally or un-intentionally), the proposed algorithm warns the user with a beep sound. As the number of warning increases, the person may reduce his/her habit and shall start avoiding touching his/her face. Also, our proposed algorithm gets improved while testing due to the Recurrent Neural network. Both, the datasets used and the algorithm being public, can be reproduced by any other researcher to solve the problem of avoidance further.

References

[1] Nicas M and Best D 2008 A study quantifying the hand-to-face contact rate and its potential application to predicting respiratory tract infection *J. Occup. Environ. Hyg.* **5** 347–52

[2] G D G, R A, L A, P M and I A 2008 A survey of knowledge, attitudes and practices towards avian influenza in an adult population of Italy BMC Infect Dis 8

[3] Kwok Y L A, Gralton J and McLaws M-L 2015 Face touching: a frequent habit that has implications for hand hygiene Am. J. Infect. Control 43 112–4

[4] Rong W, Li Z, Zhang W and Sun L 2014 An improved Canny edge detection algorithm 2014 IEEE International Conference on Mechatronics and Automation (IEEE)

[5] Viola P and Jones M J 2004 Robust real-time face detection Int. J. Comput. Vis. 57 137–54

[6] Ghanem S, Imran A and Athitsos V 2019 Analysis of hand segmentation on challenging hand over face scenario Proceedings of the 12th ACM International Conference on PErvasive Technologies Related to Assistive Environments (New York, NY, USA: ACM) (Link to Dataset: https://sites.google.com/view/sghanem/vlm-handoverface)

[7] Khan A U and Borji A 2018 Analysis of hand segmentation in the wild 2018 IEEE/CVF Conference on Computer Vision and Pattern Recognition (IEEE) (Link to Dataset: https://drive.google.com/open?id=1hHUvINGICvOGcaDgA5zMbzAIUv7ewDd3)

[8] Sudharsan B, Sundaram D, Breslin J G and Ali M I 2020 Avoid Touching Your Face: A Hand-to-face 3D Motion Dataset (COVID-away) and Trained Models for Smartwatches pp 1–9

[9] Mahmoud M, El-Kaliouby R and Goneid A 2009 Towards communicative face occlusions: Machine detection of hand-over-face gestures Lecture Notes in Computer Science (Berlin, Heidelberg: Springer Berlin Heidelberg) pp 481–90

[10] Mahrishi, M., Hiran, K. K., Meena, G., & Sharma, P. (2020). Machine Learning and Deep

Learning in Real-Time Applications. IGI Global. http://doi:10.4018/978-1-7998-3095-5

[11] Grafsgaard J F, Fulton R M, Boyer K E, Wiebe E N and Lester J C 2012 Multimodal analysis of the implicit affective channel in computer-mediated textual communication Proceedings of the 14th ACM international conference on Multimodal interaction - ICMI '12 (New York, New York, USA: ACM Press)

[12] Mahmoud M M, Baltrušaitis T and Robinson P 2014 Automatic detection of naturalistic hand-over-face gesture descriptors Proceedings of the 16th International Conference on Multimodal Interaction - ICMI '14 (New York, New York, USA: ACM Press)

[13] Naik N and Mehta M A 2018 Hand-over-Face Gesture based Facial Emotion Recognition using Deep Learning 2018 International Conference on Circuits and Systems in Digital Enterprise Technology (ICCSDET) (IEEE)

[14] Kim P 2017 Convolutional Neural Network MATLAB Deep Learning (Berkeley, CA: Apress) pp 121–47

[15] Albelwi S and Mahmood A 2017 A framework for designing the architectures of deep convolutional neural networks Entropy (Basel) 19 242

ICETCE 2021 IOP Publishing
IOP Conf. Series: Materials Science and Engineering 1131 (2021) 012007 doi:10.1088/1757-899X/1131/1/012007

Impact of Hidden Dense Layers in Convolutional Neural Network to enhance Performance of Classification Model

Helen Josephine V L[1], A.P.Nirmala[2], Vijaya Lakshmi Alluri[3]

[1]Associate Professor, Department of Computer Applications, CMR Institute of Technology, Bengaluru, India
[2]Associate Professor, New Horizon College of Engineering, Bangalore
[3]Research Scholar, Acharya Nagarjuna University, Andhra Pradesh

helenjose.cbe@gmail.com

Abstract. Education and Health care sectors are two predominant areas where societal growth is expected through innovation and technology development. Machine Learning and Deep learning classification models have been entertained in predicting, detecting, and diagnosing major diseases in the early stage. In this research paper, we have analyzed the impact of hidden dense layers in the Convolution neural network to improve the performance of the classification model. Three different classification deep network models have been constructed, analyzed and the result was tested with a diabetes dataset. Results concluded that the more layers with deeper the network better was the classification performance. The classification model with six hidden dense layers outperforms all other less number of hidden dense layers.

Keywords: Deep Neural Network, Deep Learning, Diabetes Prediction, Classification

1. Introduction

Diabetes is a continual disease which takes place when the internal organ pancreas does not produce adequate hormone insulin or the body which is not the possible to successfully consume the segregated hormone-insulin. Uncontrolled or untreated diabetes leads to high blood sugar in the blood and every organ including the heart and vision will be function properly. The medical report states that diabetes patients' growth rate increased rapidly, from 26 million (1990) to 65 million (2016). Early detection of the disease will help the patient to regulate his food habit and undergo proper medication.

Traditional computer programs are expecting input data and algorithms to be fed to provide output. Whereas, machine learning methods are receiving input data as well as label data, which in turn will return the model or algorithm. Deep learning methods are the subset of machine learning, which have several combinations of an artificial neuron or multilayer perceptrons connected in different ways to operate several activation functions[1]. These kinds of deep learning architectures have been predominantly used to create smarter applications in all domains which are primarily based on the artificial natural network. It entertains multi-layers of nonlinear processing for feature extraction and transformation. The main idea of deep learning is exhibited in behavioral patterns in the different layers of neurons in the neocortex of the human brain[2][3].

Content from this work may be used under the terms of the Creative Commons Attribution 3.0 licence. Any further distribution of this work must maintain attribution to the author(s) and the title of the work, journal citation and DOI.
Published under licence by IOP Publishing Ltd

In our present work, we constructed three different classification deep models to analyze and predict diabetes. We have achieved a high accuracy value of 98.6% for 6 dense layers with various numbers of neurons in each layer. It was compared with the other two classification deep models with 3 and 4 dense layers.

This research paper constructed three different classification deep models to predict the early detections of diabetes. Section II summarizes the existing research in this domain. Section III illustrates the Methodology to construct a deep model. Section IV discusses the datasets and experiment setup. The results were analyzed and discussed in Section V. Eventually, section VI concludes the research paper.

2. Literature Survey

Zhou et.al.,[4] utilizes MSN – Medical-Social-Network based approach to evaluate the extrapolative value of pathological factors to detects early Heart failure detection. The author makes use of EHR - Electronic Health Records which has been collected through the Heart Carer project to calculate the likeness of pathological risk factors. By using these values help to construct the weighted and un-weighted MSN – Medical-Social-Network. GD Group-division algorithm has been used to further split the MSN into two groups' namely high -risk heart failure and low -risk heart failure groups.

Dilshad et al.,[5] constructed a classification model which is based (Conv-LSTM) Convolution Long Short-Term Memory (Conv-LSTM) that was not functional yet in this observe. Researchers utilized PIDD (Pima Indians Diabetes Database) to compare the state of deep learning algorithms, Convolution-Neural-Network (CNN), Traditional LSTM (T-LSTM), and CNN-LSTM.

In one research article Author[6] entertained flattened networks which comprised of many sequences with filters (1D) to acquire similar performance like CNN. The model has been tested with the various dataset and concluded that the flattened layer is a very good alternate for the 3D filters without compromising the accuracy. The proposed model has one more advantage that one it has been trained on, does not need post-processing or physical tuning.

Numerous research has been initiated in a deep neural network with CNN[7],[8]. Liver and Brain Tumors disease has been classified by CNN and Discrete Wavelet Transform and LSTM Network, [9]. Deep CNN LSTM model for forecasting in smart cities

3. Methodology

Convolution neural networks are not only used for image classification but also all other types of dataset. In this session, will see various types of convolution.

3.1 Types of convolution

There will be different type's convolution which has been used in the Convolution Neural network. The primary types are convolution, dilated convolution, and transpose convolution. These are often used in deep network and also in other networks which contains decoder and encoder kind of architecture [10].

A) Naïve Convolution

Naive Convolution is having three important parameters, Kernel size – height and width, Stride – height and width, Padding. The dimension of the output feature map is governed by this parameter. The output width and output height are calculated by equation No (1) and equation No. (2). Kernal width will be subtracted from input width and it will be added with 2 times padding ad this value will be divided with stride width and the entire result will be added with one.

$$O_w = \frac{I_w - K_w + 2p}{S_w} + 1 \tag{1}$$

$$O_h = \frac{I_h - K_h + 2p}{S_h} + 1 \tag{2}$$

B) Dilated convolution

Dilated is also known as Atrous convolution. In this type of convolution, the kernel is filled with zero to see a larger receptive field. It has an additional parameter called "Dilation Rate". If the dilation rate is set 2 for any 3x3 convolution then it would produce the same result as naïve 5x5 convolution with all other parameters.

C) Transposed Convolution

The idea behind the transposed convolution is to assist in increasing the output feature map size. It helped in encoder-decoder networks to increase the spatial dimension of the feature map. In transposed convolution rebuilt the original optical evolution and executes a convolution. The input image will be padded before the convolution operation.

3.2 Classification deep model

Single or several convolution layers that regularly come with sub-sampling stage consist of Convolution neural networks. These layers are then connected with several fully-connected layers. The complete architecture will be similar to standard multilayer neural networks. The image data set will be primarily used on CNN algorithms. But CNN algorithms will also be used in the numerical dataset to bring out high accuracy and performance.

Keras is one of the open-source libraries which has python interface for artificial neural network, especially for the Tensorflow open source library. It provides four important pre-build layers namely core layers, convolution layers, pooling layers, and recurrent layers. By using these layers, we would create any complex neural network easily. It contains sequential and functional models. The word sequential means linear combinations. Sequential model is easy, smallest as well as can signify nearly all available neural networks. The dense layer is the general and often used layer that contains a deeply connected neural network layer. It can be used as it means a hidden layer that is associated to each and every one node in the next layer. Any dense layer will perform the below operation

$$Output = activation(dot(intput, kernel) + bias) \tag{3}$$

The sequential model contains accurately one input tensor and one output tensor in each layer, which is appropriate for **a simple stack of layers.** When a sequential model has been built, it acts like a layer that contains an input and output attribute. In the sequential model, these attribute to create a model which excerpts the output of all intermediate layers. A Sequential model is not suitable when the model has several inputs and several outputs and when there is an expectation for non-linear topology.

Hidden dense layers place a very vital role in the neural network. In this research work, we have constructed three sequential models. Each has its dense layers and the various number of neurons in each layer. The second sequential model network architecture is presented in Figure. 1 which contains input layers. The dataset contains numerical data that has to be normalized in the pre-processing step. Seven attributes were normalized and concatenated. It is followed by several dense layers. The second classification deep model contains 4 dense layers, each layer contains 32, 24, 12, 8 neurons respectively. In the third classification, deep models are comprised of 6 dense layers and each layer has a different number of neurons. The irst dense layer has 64 neurons , second dense layer contains 64 neurons, third dense layer contains 32 neurons, followed by fourth layer which has 24 neurons. The fifth and sixth layer has 12 and 8 respectively. Figure. 2 depicts the architecture diagram.

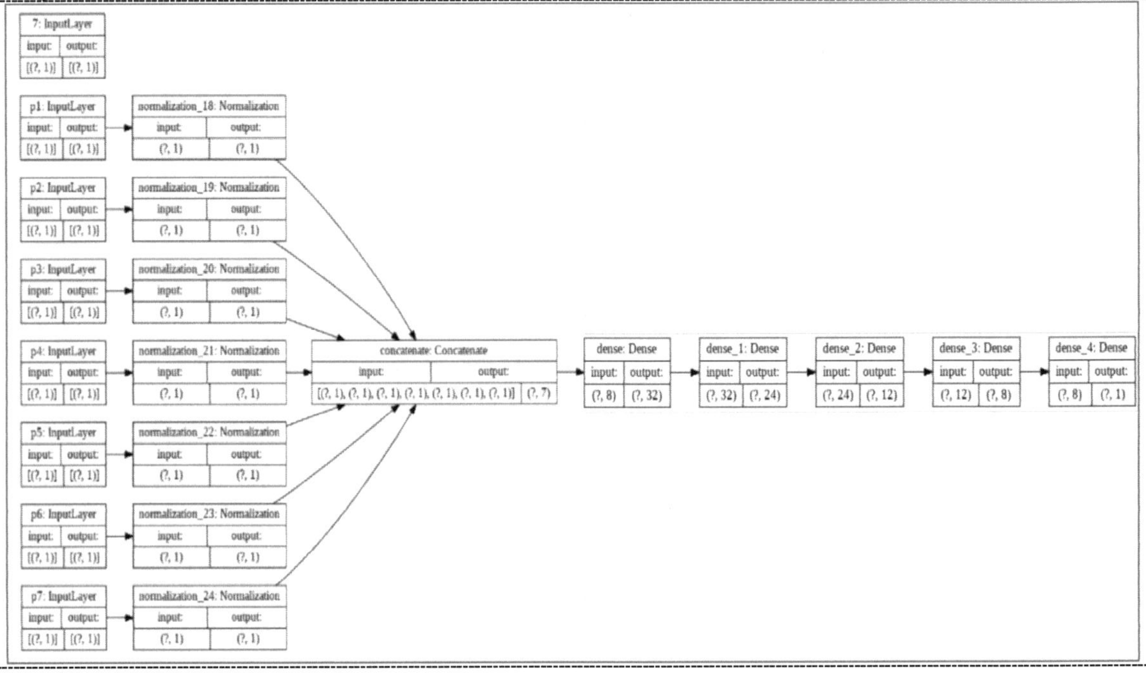

Figure 1. Classification deep model with 4 layers

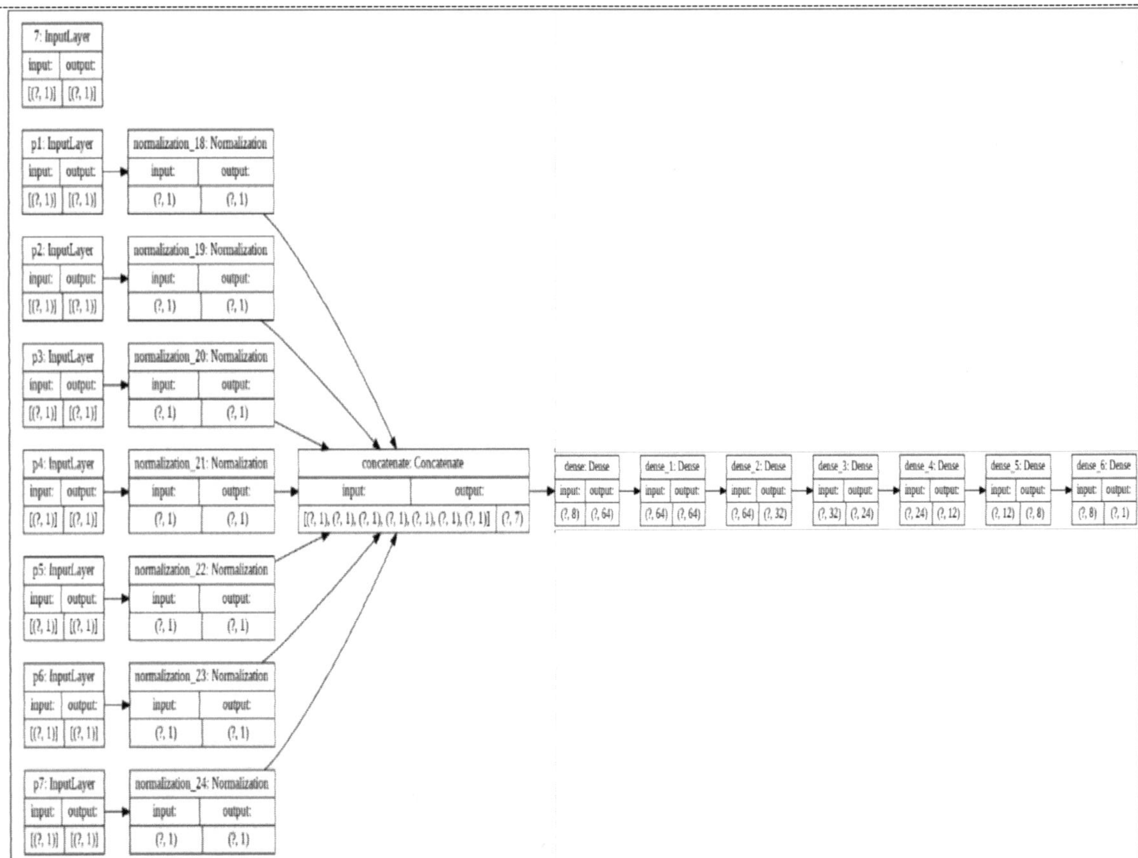

IOP Conf. Series: Materials Science and Engineering 1131 (2021) 012007 doi:10.1088/1757-899X/1131/1/012007

Figure 2. Classification deep model with 6 layers

4. Dataset & Experiment Setup

The performance of the Deep-Convolution Neural Network model was estimated through a famous diabetes database called Pima Indian Diabetes Database (PIDD) and it was downloaded from the UC Irvine Machine Learning Repository. This dataset includes 768 instances. Dataset has 9 important attributes to identify diabetic disease. Table 1 shows the attributes of the dataset.

Table 1. PIDD Dataset Attributes

Attribute No with observation	Description
1-NTP	Number of times pregnant
2-PGC	Plasma glucose concentration a 2 hours in an oral
3-DBP	Diastolic blood pressure (mm Hg)
4-TSFT	Triceps skin fold thickness (mm)
5-H2SI	Hour serum insulin (mu U/ml)
6-BMI	Body mass index (weight in kg/(height in m)^2)
7-DPF	Diabetes pedigree function
8-AGE	Patient's Age
9-Lable	Class 0 or Class 1 - Diagnosis of diabetes with

5. Experiment Result and Analysis

To speed up feed-forward execution, we have increased hidden layers and engaged a flattened convolution neural network. Table No. 2 shows the overall performance of three different classifications deep models with the various number of hidden layers.

Table 2. Performance of the classification deep model

Total Number of dense layers	No of the neurons in each layer	Accuracy	Precision	Recall	F1 Score
3	12,12,8	0.832	0.762	0.754	0.758
4	32,24,12,8	0.895	0.819	0.896	0.856
6	64,64,32,24,12,8	0.986	0.964	0.996	0.986

Figure 3 and Figure 4 depicts the loss and accuracy graph for training and validation for the second and third classification deep model. Batch size has been set as 32 and the total number of epochs is 500 for all the classification deep models.

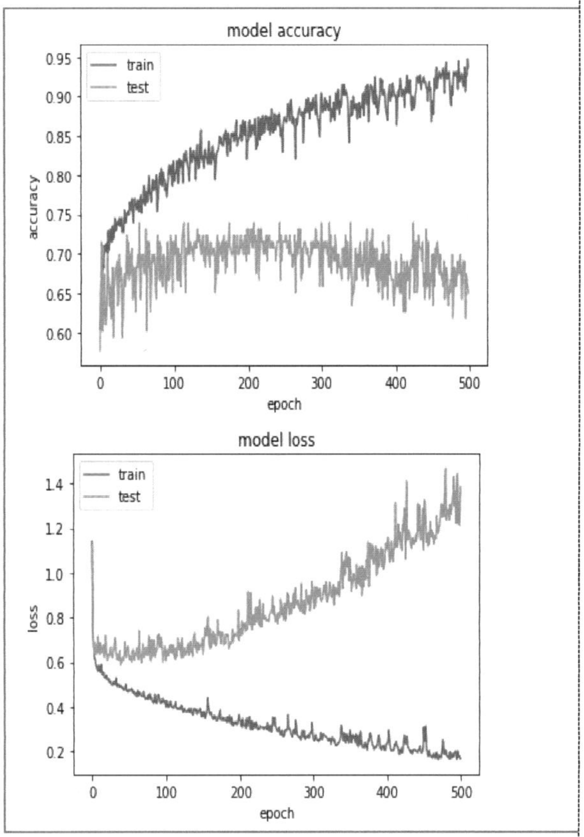

Figure 3. Loss and accuracy graph for training and validation - Classification deep Model with 4 hidden layers

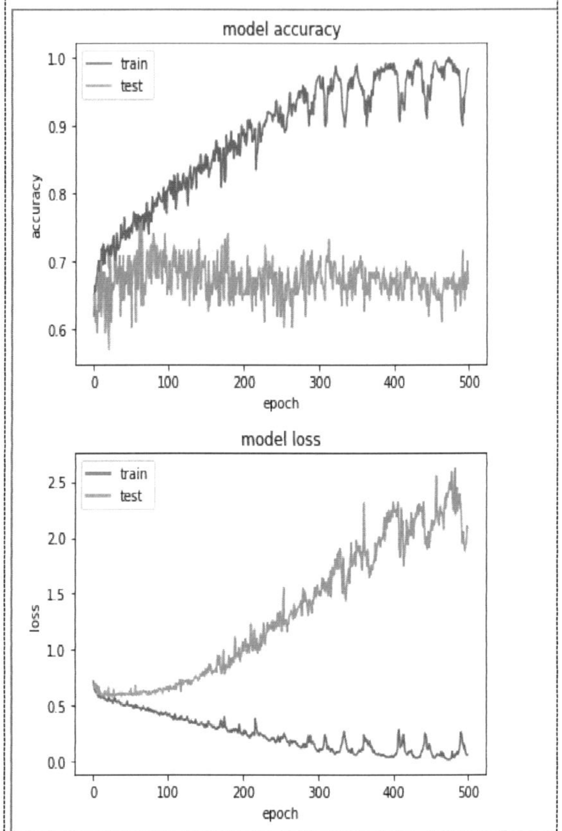

Figure 4. Loss and accuracy graph for training and validation - Classification deep Model with 6 hidden Layers

The first classification deep model contains 3 hidden layers with 12,12 and 8 neurons in each layer respectively. Classification performance is measured with four important parameters namely Accuracy, Precision, Recall, and F1 Score. The result shows that accuracy 83%, precision 76%, recall 75%, and F1 Score 75%. The second classification deep model comprised of 4 hidden layers with 32, 24, 12, and 8 neurons. Performance has been increased 7%, precision by 8%, recall by 15%, and F1 Score by 10%. The final test was conducted with 6 hidden layers classification deep model. Each layer has 64, 64,32,24,12 and 8 neurons respectively. The confusion matrix presented for 4 dense and 6 dense layers with various numbers of neurons in each layer in Figure. 4 and Figure 5 respectively. When there is an increase in the number of layers and number of neurons in each layer, CNN outperforms compared to the other classification algorithms. The final experiment outcomes produce 99% accuracy, 96% precision, 100% Recall, and 98% F1 Score. Classification Accuracy has been increased from 83% to 99% from 3 dense layers to 6 dense layers. If we further increase the hidden layer the performance will be is saturated since it reaches the highest learning point.

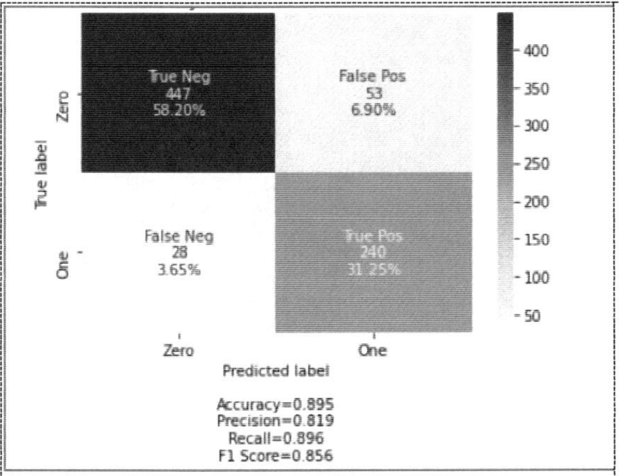

Figure 5. Confusion Matrix for Classification deep Model with 4 hidden layers

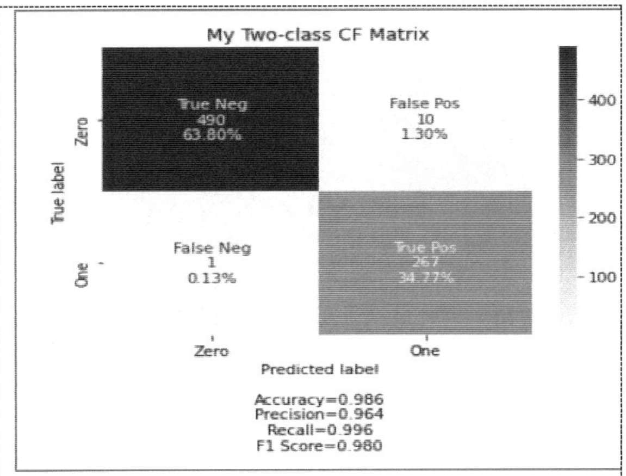

Figure 6. Confusion Matrix for Classification deep Model with 6 hidden layers

6. Conclusion

In this research paper, we have constructed three different classification deep models to predict type 2 diabetes. The models were analyzed based on the classification performance metrics accuracy, precision, recall, and F1 -score. The Results concluded that the deeper the model which has more dense layers with more neurons in each hidden layer outputs the less the number of hidden dense layers. The more hidden layers with the deeper the network better were the classification performance.

References

[1] Krizhevsky, A., Sutskever, I., and Hinton, G. E.. 2012, *Imagenet classification with deep convolutional neural networks.* Advances in neural information processing systems, pp. 1097– 1105.

[2] Montufar GF, Pascanu R, Cho K, Bengio Y., 2014, On the number of linear regions of deep neural networks. In: *Advances in Neural Information Processing Systems*. Red Hook, NY: Curran Associates, Inc. pp. 2924–32.

[3] Mahrishi M., Morwal S. (2020) "Index Point Detection and Semantic Indexing of Videos—A Comparative Review". In: Pant M., Kumar Sharma T., Arya R., Sahana B., Zolfagharinia H. (eds) Soft Computing: Theories and Applications. Advances in Intelligent Systems and Computing, vol 1154. Springer, Singapore.

[4] C. Zou, A. Li, A. Hou, Z. Zhang, Z. Zhang, F. Wang, Modeling methodology for early warning of chronic heart failure based on real medical big data. Expert Syst. Appl., Published online. https://doi.org/10.1016/j.eswa.2020.113361

[5] Rahman, Md & Saha, Indrajit & Islam, Dilshad & Mukti, Rokeya Jahan. (2020). A Deep Learning Approach based on Convolutional LSTM for Detecting Diabetes. Computational Biology and Chemistry. 88. 107329. 10.1016/j.compbiolchem.2020.107329.

[6] Esteva, A. Robicquet, B. Ramsundar, V. Kuleshov, M. DePristo, K. Chou, J. Dean, 2019, *A guide to deep learning in healthcare.* Nat. Med. 25(1), 24–29

[7] Zhou, H., Myrzashova, R. & Zheng, R., 2020, Diabetes prediction model based on an enhanced deep neural network. *J Wireless Com Network* , 148. https://doi.org/10.1186/s13638-020-01765-7.

[8] Kutlu H, Avcı E., 2019 *A Novel Method for Classifying Liver and Brain Tumors Using Convolutional Neural Networks, Discrete Wavelet Transform and Long Short-Term Memory Networks. Sensors (Basel, Switzerland).* Apr;19(9). DOI: 10.3390/s19091992.

[9] Huang CJ, Kuo PH. 2018, *A Deep CNN-LSTM Model for Particulate Matter (PM2.5) Forecasting in Smart Cities. Sensors (Basel)* Jul 10;18(7):2220. doi: 10.3390/s18072220. PMID: 29996546; PMCID: PMC6069282.

[10] Yamashita, R., Nishio, M., Do, R.K.G. et al. Convolutional neural networks, 2018, *An overview and application in radiology. Insights Imaging* Vol.9, pp.611–629. https://doi.org/10.1007/s13244-018-0639-9.

IOP Conf. Series: Materials Science and Engineering 1131 (2021) 012008 doi:10.1088/1757-899X/1131/1/012008

Statistical Measures for Trending Video Analysis using knowledge pre-processing and preliminary knowledge Analysis

Rahul Pradhan, Akansha Saxena, Divyanshi Bansal and Ananya Jain

GLA University, Mathura

akansha.saxena0211@gmail.com

Abstract. YouTube has an algorithm that decides which videos will appear in the Trending Videos List. Some are more likely to be on the trending list but some videos which were never estimated to be on-trend are seen, it targets viewers' attention over a short span. On Youtube, trending video lists are not personalized according to a person's views, likes, or dislikes just like what the recommendation system does, rather its algorithm targets a collective group of people, the same Trending video page is displayed to everyone in a country. In India, the Trending page of. Like various recommendation systems, YouTube also gives you suggestions on your Home page based on your previously watched content. Recommendation systems are so common that we are using it without even knowing it, it makes the user experience much better. Common Examples are Amazon, NetFlix, and now Spotify uses it to suggest to us the music we might love to listen to. The trending videos page updates itself every few minutes (15 minutes). After every update, videos may move up, down, or stay in the same position in the list. Here in this Paper, we used statistical skills to analyze what are the reasons for a video to be on trends. In this paper, we have worked on a dataset for the year 2020. For data collection, we used a Web Scraper script, that's not part of our work. Based on EDA we can say what is the exact reason for a video to appear on the trending page.

1. Introduction

There is no other platform as popular as YouTube when it comes to video content. The video uploading rate is very high and billions of views are there in a single day. There are times when you search for motivational videos or some videos to save your life during exams, talking about tonnes of educational content available.

Try to keep some points in your mind while uploading your very own video:

- Videos must be appealing to a wide range of viewers
- You must ensure that videos are not at all misleading
- Capturing the breadth of what's happening in the world
- Work on unique ideas that might bring more and more viewers to your video.
- A video might appear on a trending page due to some of these reasons:
  - View count
  - Rapid growth in views with time
  - Where views are coming from (including outside of YouTube like Google search, Facebook, etc.)

Content from this work may be used under the terms of the Creative Commons Attribution 3.0 licence. Any further distribution of this work must maintain attribution to the author(s) and the title of the work, journal citation and DOI.
Published under licence by IOP Publishing Ltd

The exponential increase in the amount of data that is generated every year is demanding us to look at it as data is becoming more and more critical. The analysis includes simple queries, statistical analysis, and data mining. Having so much content available, social media streams what people are paying attention to and what is the hype about certain topics it becomes a need to have data analysis's conclusions. Correctly utilizing trending topics requires a better understanding of their various characteristics in different social media streams and it requires statistical skills to draw some conclusions. As part of our Project, we have collected datasets through YouTube data API which provides a systematic collection of public standard feeds and statistics related to videos and users. We aim to produce a scientific knowledge preprocessing analysis operating solely with the dataset INDIA Videos. This step is important for all data processing exercises and that we wish to emphasize it. We will first grasp key knowledge attributes, like missing values, distinctive counts, outliers, and time-series trends.

This Paper aims to function as a help to find trends in the video which will help a channel to grow and will get to know about what kind of videos are coming in trends. The focus is only on the INDIA videos dataset that isn't too huge by big-data standards. This knowledge set contains solely YouTube data like video, image, audio, or giant text documents. Still, we are going to proceed with knowledge preprocessing and preliminary knowledge Analysis (EDA) as if this were a very huge dataset, using techniques that might be utilized in rather more difficult knowledge manning exercises.[1]

The data used in this analysis were retrieved using the Youtube API. A script was to fetch data from YouTube data API about trending videos of each day. The script in which the youtube API key is passed after making an account at google console for YouTube and then processes the data received from YouTube and stores them in text files. The scraping script is not part of our work, only worked on the EDA part [2].

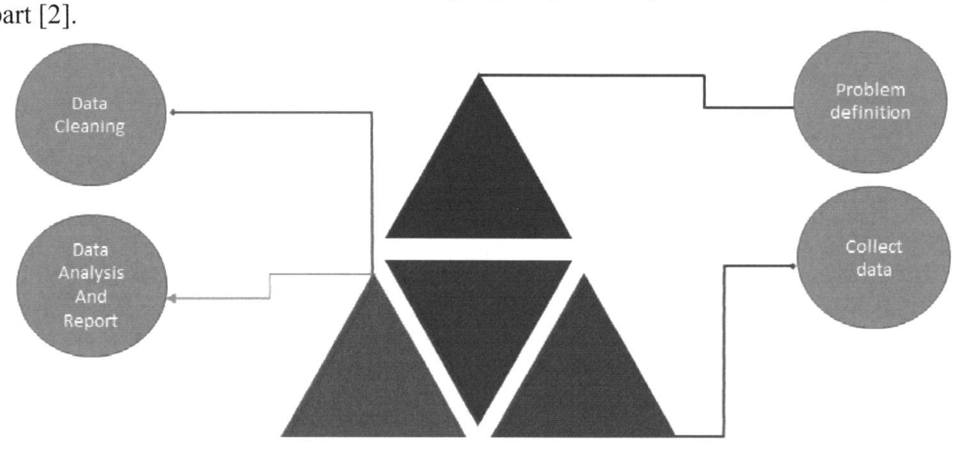

Figure 1. The workflow of a data analysis project

2. Related Work

(Barjasteh et al., 2014) [3] mentioned that there is no such statistical analysis on the trends of YouTube videos. This statistical analysis hasn't received the attention which it should have received seeing the significance of this analysis. With the years passing by YouTube has become the best place for advertising your brand, people might show their response be it positive or negative, or maybe neutral. They analyzed and evaluated 4,000 trending videos that were collected for about 2 months using YouTube API. Dataset itself describes the number of views, comments, time duration, its category. It includes a comparison between trending and non-trending videos, from which they were able to draw some statistical attributes out of trending videos. Analyzed the profile of uploaders. It might not seem so obvious but it can influence the video. Basically what they analyzed were the subscriber's count and other uploaders' stats. They also analyzed the directional-relationship between the trending videos and the trending videos that became popular. They used Granger Causality(GC) with significance testing to measure the directional relationship. [1]

IOP Conf. Series: Materials Science and Engineering 1131 (2021) 012008 doi:10.1088/1757-899X/1131/1/012008

(Bärtl.,2018) [4] aimed at how video consumption evolved over the course of time. Through the results of an analysis, it was evident that older videos have a higher probability to gain large viewership than the young channels, it all depends on the topic they select for making their video. Since this analysis was done on a dataset of 10 years, as a result, they were able to observe the gradual growth of various channels. So what they were focusing on was how YouTube evolved in terms of channels, uploads, and views. Identified what were the factors that were responsible for channels' growth in terms of viewers', subscribers' count. [4]

YouTube videos are uploading rapidly in real-time. Users can share, comment, and rate (like or dislike) videos [6].

(Bhuiyan et al.,2017) [5] worked on NLP processing based sentiment analysis on comments by the users., which helps in finding out the emotion of the video. To date, YouTube is the best platform for the compilation of video content on the web. The versatility and variety of content attract users across the globe. Sometimes it also happens that we get nothing we are looking for. Happens mostly when we are searching for a particular movie instead what we get is some movie's cut scenes or its trailer. This kind of scenario occurs frequently due to likes that video has got.[3] With all these things happening on a platform, there is some kind of interest in analyzing the rich content shared on YouTube. It has been seen that comments might be a good source to perceive the video content i.e its quality, the relevance of the topic, and gives us the idea of what kind of message the video is giving. [3]

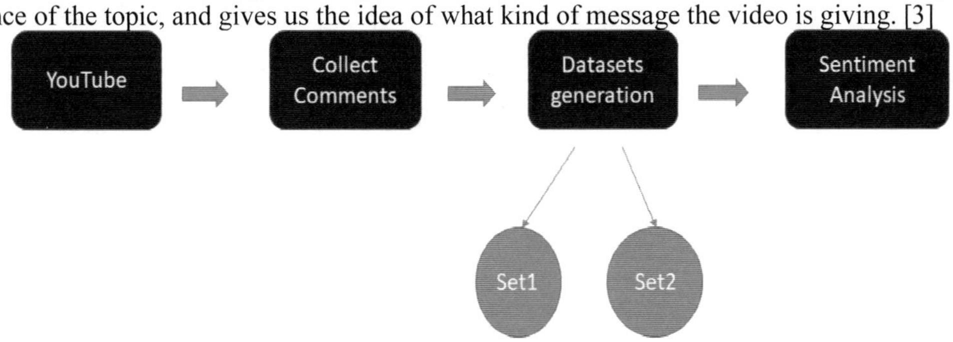

Figure 2. Workflow for sentiment analysis

(Amudha et al.) performed Data Analysis on 5 Million video records from YouTube API, a huge dataset so can't be analyzed by traditional databases and needs tools for parallel processing. The data size was 603 MB. They used Anaconda Jupyter Notebook. Challenge was analyzing such large-scale unstructured data of video content.[2]

Some of the Big data tools are Hadoop, MongoDB, Cassandra, Drill, Elasticsearch, HCatalog, Oozie, Storm. Computers work in 2 ways i.e Real-Time and Batch Processing

Real-Time as the name suggests data computation and processing needs to be performed on the data, arriving in real-time. The best example is when you withdraw money from an ATM using your ATM card. The bank needs to process your request almost instantaneously. Credit card transaction is also its application, the decision of whether the transaction is allowed or declined is made while the transaction is still pending. Batch Processing, transactional data is collected over a while and then data is processed and output is stored. One of the best examples of batch processing is Payroll Processing.

(Stefan Siersdorfer et al., 2010) did an in-depth study of comment rating behavior of around 6.1 million comments on 67290 videos on YouTube. They also analyzed what were the exact dependencies between topics, views, comments. They also studied the sentiment of comment and its impact on the video rating, for this they used SentiWord vet thesaurus. They created their test collection by formulating queries from Google's Zeitgeist archive from 2001 to 2007 For each video around 500 comments were gathered and they all were analyzed for comment rating.[6]

There are various techniques for sentiment analysis which are performed on user comments and for this SentiWordNet is used majorly . In sentiment analysis we deal with mainly 3 issues: Classification of events , Sentiment polarity detection, and prediction of youtube comments . Detection of sentiment polarity could be Lexicon based , negation based or SMAPD. Evaluation is done on the basis of precision and recall . If you make a search on your favourite search engine , what it does is it returns 30 pages (say) and only 20 pages are relevant . Then its precision is 20/30. After that classification is done using SVM . Classification is done on a dataset of user comments that is divided in two parts i.e Test data and Training data.[7]

Figure 3. Sentiments

3. Proposed Approach

There are a series of steps involved for a conclusion before getting results of the analysis on data:

- Data Collection
- Data Preprocessing
- Data Visualization
- Build Model

The analysis is done on the basis of likes, dislikes, title, description, comments.

Based on this we analyzed what was the reason for a video to be trending. We extracted data for performing data analysis on YouTube videos to get into the latest trends. Since YouTube data grows rapidly with a great speed, that is why there is a huge demand to store and process it.

Before data insights, we would remove missing values, distinctive counts, outliers, and time-series trends that are a part of the data cleaning process.[8]

3.1 Problem Definition

This paper is based on the analysis of Trending Videos on YouTube, Nowadays, every 10th person has a YouTube channel and is uploading video content. Some videos become popular which leads to the growth of a channel but for some channels, it takes too long to grow, it all depends on what topic that channel is covering through its videos.

There are various types of content available on YouTube:

- Entertainment
- Sports
- Music, Genres, Albums
- Daily Soap
- Trailers
- Workout videos of fitness experts
- Cooking Videos
- Educational videos

- Romantic videos
- Spiritual videos

And it becomes interesting as well as significant enough to explore the insights that why a video is trending, what is it depending on; whether it is the category or the particular title which is catching users' attention. We want to dig deep to analyze the relationship between various characteristics of trending videos. Some videos are controversial due to the content it has. Like a few days back in 2020 there were lots and lots of videos on the Sushant Singh Rajput case. Many YouTubers made videos on it as the title of their videos. We want to analyze how cultural divergence is affecting likes, comments.

3.2 Dataset Description

We created a dataset named "Trending Video Analysis" which was built using YouTube API Services, for this, we have merged several CSV files which were having data of a couple of weeks on trending videos from YouTube for India. Dataset is having more than 1000 observations, having information such as video title, category, publish time, view count, likes, dislikes, comments, description, channel id, video id, etc. Before working on this we had a few questions like what are the characteristics which trending videos possess. We measured user engagement through these characteristics.

3.3 Plan

The analysis was performed using a Jupyter Notebook
We examined view counts on a video. We also examined likes and dislikes for a video that is trending to get to know what is the viewers' attitude towards that type of video. We also get to know what were the common words found in the title of a trending video like: "official", "trailer", "music", etc.

3.4 Visualization

We used: Seaborn, Numpy, Pandas. We also used NLP Libraries. We performed this analysis using Python and Python libraries like Pandas, Matplotlib, NLTK, WordCloud.

4. Analysis Results

Through the results of the analysis on our dataset, we are able to see that the trending video having the highest view count is by BTS band names 'Dynamite'OfficialMV. It has around 16124714 likes and 757828 dislikes.
Table below is the information about the trending video having the highest view.

Table 1 Video Attributes

Video Id	gdZLi9oWNZg
Title	Dynamite Official MV
Channel Title	Big Hit Labels
Likes	16124714
Dislikes	757828
View Count	254687524
Comments Count	6255536

4.1 Most common words in the trending videos

While analyzing the dataset we also found what were the most common words in the trending videos.

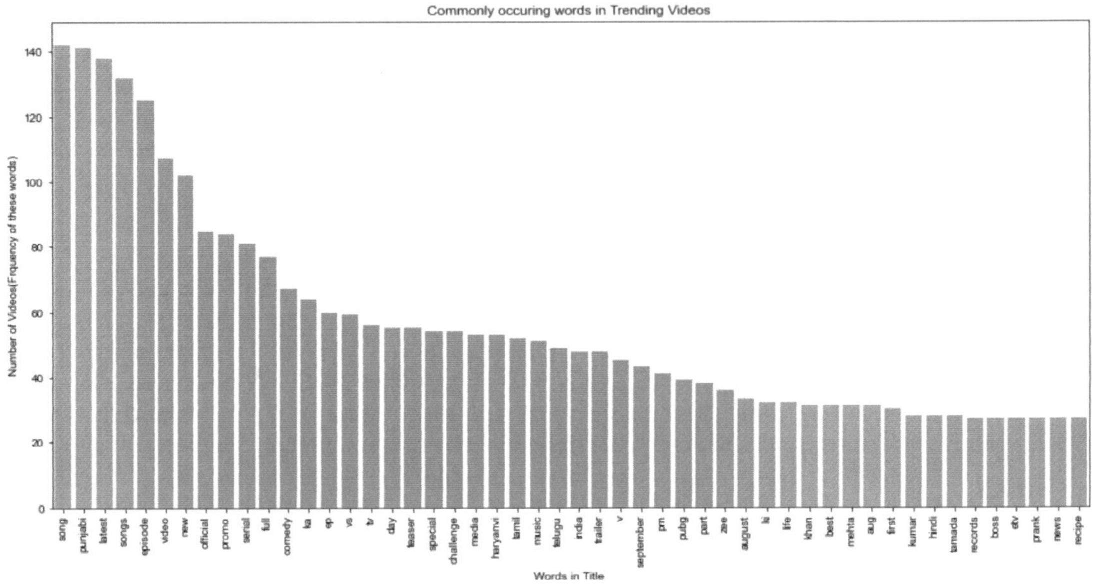

Figure 4. Commonly Occurring Words in Trending Videos

Figure 5. WordCloud for commonly occurring words in trending videos

Figure above is a representation of Fig 4 in Word Cloud which represents the words on X axis(Commonly occurring words in Trending Videos)

4.2 Channel titles

We also got to know through our analysis that what are most Channel titles of trending videos

Those were:

- Technical Guruji

- ZEE5
- Taarak Mehta Ka Ooltah Chashmah
- Speed Records
- SAB TV
- Desi Music Factory
- Colors TV
- SET India
- Anysaa

There were some more of them, but here we listed only those which we all are aware of, to just get a glimpse of what are the channel titles which became trending so that you instantaneously get to know what would be the type of content. You can see it through the graph below.

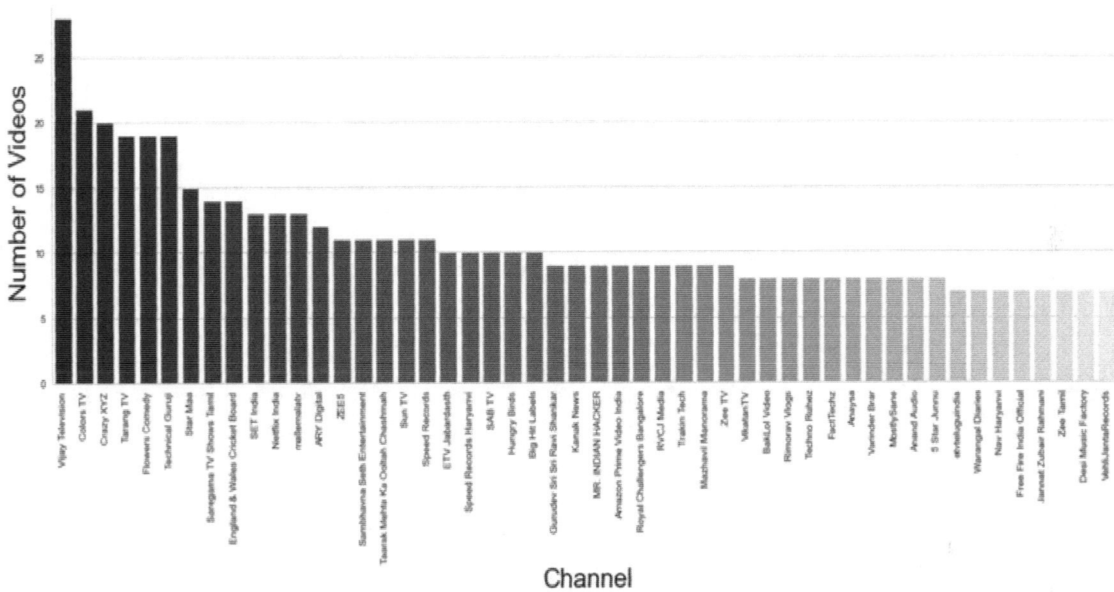

Figure 6. Common Channel titles

Image above is small so the channel titles on X axis are mentioned in the bulleted list.

Figure 7. Word cloud for common channel title

4.3 Views on Trending Videos

We also get to know what were the Views on Trending Videos

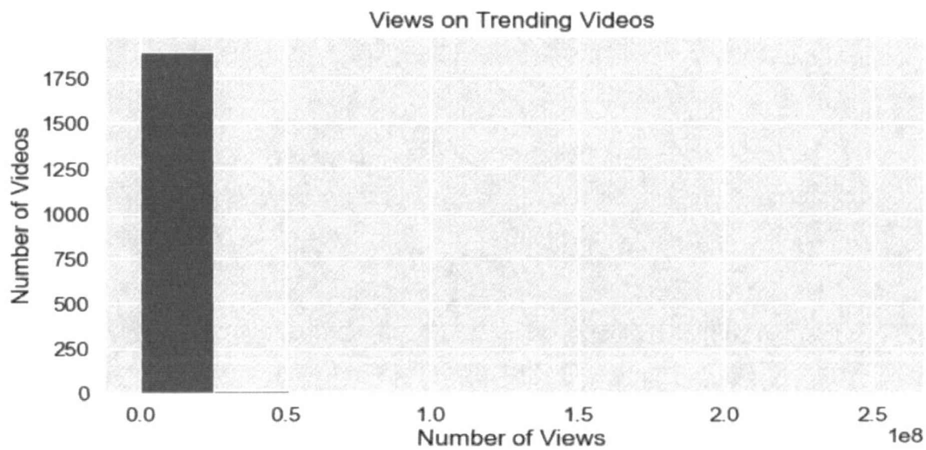

Figure 8. Views on Trending Videos

Through the figure above we can observe that the majority of videos have views less than 5 Million.

4.4 Videos with Less than 5 Million views

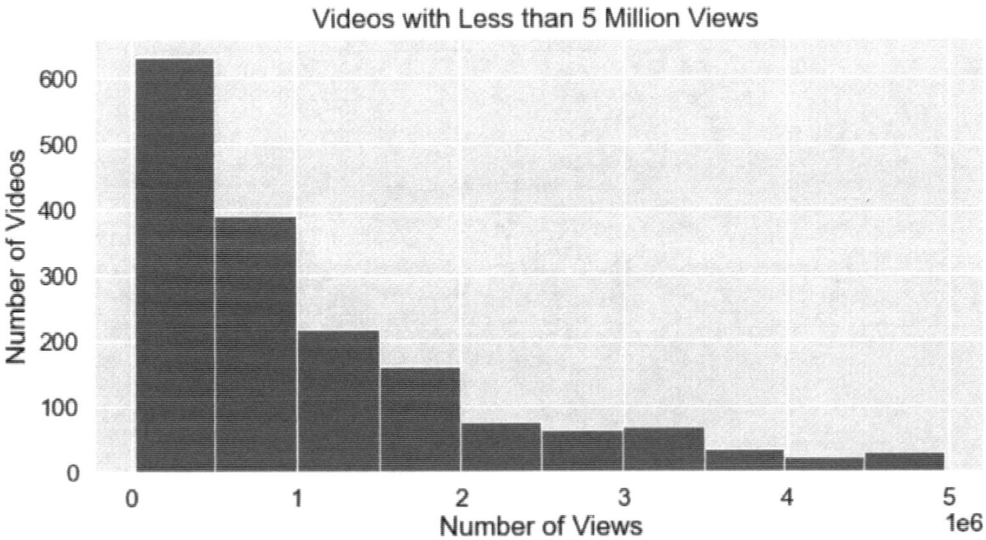

Figure 9. Views of Trending Videos

(In Figure 9 data has been selected in order to have videos lying in the category of views less than 5 million)

From the figure above we can see that most of the videos have less than 1 Million views

4.5 Likes on Trending Videos

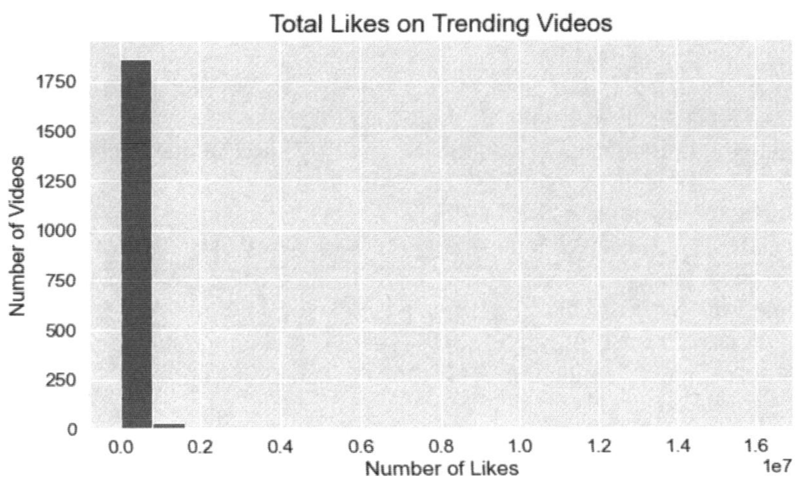

Figure 10. LIkes on Videos

From the above figure, we can get to know that most of the trending videos have liked by less than 5 Million.

4.6 Likes on Trending Videos Below 20,000

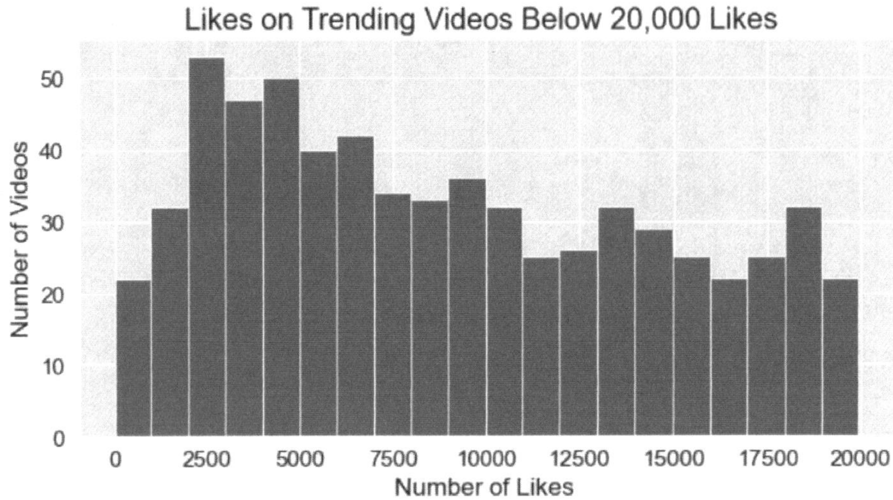

Figure 11. Likes on Videos

5. Conclusion

From the results of our analysis, we can conclude that we found some obvious channels on the trending list. Those were: Technical Guruji, ZEE5, Tarak Mehta Ka Ooltak Chashmah, Speed Records, SAB TV, DesiMusic Factory. We found that most of the videos have less than 1 Million views. Common words found in the title of the trending videos were: "Official", "Trailer", "new", "Punjabi", "song". The trending video having the highest view count is by a BTS band named "Dynamite'OfficialMV".

References

[1] Barjasteh, I., Liu, Y., & Radha, H. (2014). Trending videos: Measurement and analysis. Retrieved from http://arxiv.org/abs/1409.7733

[2] Amudha, S., Niveditha, V. R., Kumar, P. R., Revathi, M., & Radha, S. (n.d.). Youtube Trending Video Metadata Analysis Using Machine Learning.

[3] Bhuiyan, H., Ara, J., Bardhan, R., & Islam, M. R. (2017). Retrieving YouTube video by sentiment analysis on user comments. 2017 IEEE International Conference on Signal and Image Processing Applications (ICSIPA). IEEE.

[4] Bärtl, M. (2018) 'YouTube channels, uploads, and views: A statistical analysis of the past 10 years', Convergence, 24(1), pp. 16–32. DOI: 10.1177/1354856517736979.

[5] Stefan Siersdorfer, Sergiu Chelaru, Wolfgang Nejdl, and Jose San Pedro. (2010). How useful are your comments? analyzing and predicting youtube comments and comment ratings. Association for Computing Machinery, New York, NY, USA, 891–900. DOI:https://doi.org/10.1145/1772690.1772781

[6] Mahrishi, M., Hiran, K. K., Meena, G., & Sharma, P. (2020). Machine Learning and Deep Learning in Real-Time Applications. IGI Global.

[7] Siersdorfer, S., Chelaru, S., Nejdl, W., & San Pedro, J. (2010). How useful are your comments?: Analyzing and predicting youtube comments and comment ratings. Proceedings of the 19th International Conference on World Wide Web - WWW '10. New York, New York, USA: ACM Press.

[8] Mohana Prabha, B. M., & Ramya, R. P. (Eds.). (n.d.). Predicting the Popularity of Trending Videos in Youtube Using Sentiment analysis.

Bitcoin Price Alert and Prediction System using various Models

Ashutosh Shankhdhar[1], Akhilesh Kumar Singh[2], Suryansh Naugraiya[3] and Prathmesh Kumar Saini[4]

[1,2,3,4] Department of Computer Engineering and Applications GLA University, Mathura, 281406, India

E- mail :
[1]ashutosh.shankhdhar@gla.ac.in,[2]akhilesh.kumar@gla.ac.in,[3]suryansh.naugraiya_cs18@gla.ac.in, [4]prathmesh.saini_cs18@gla.ac.in

Abstract. From last many years it has been a trend to invest in cryptocurrency especially (Bitcoin) because it is one of the most popular and decentralized digital currency. However, its prices keep on fluctuating very much that makes it difficult to predict. So, our research aim is to find the less time consuming and accurate model for the prediction of Bitcoin price from different machine learning models like (Multivariate Linear Regression, Theil-Sen Regression, Huber Regression) and deep learning algorithms like (LSTM, GRU). The dataset that we will use for our prediction purpose will be stored in MongoDB (Big-Data Tool) because it consists of huge data points. We have also implemented IOT in our system to create an alert system, which alerts user when the value of bitcoin price reaches a threshold value.

Keywords— Big Data, IOT, Machine Learning, Deep Learning, Price Prediction, Price Alert, Bitcoin, Cryptocurrency.

1. Introduction

In the modern era, cryptocurrency is a trendsetter. Within the finance sector, with the passing of time Bitcoin has emerged as one of the popular digital decentralized cryptocurrencies. It was introduced by Satoshi Nakamoto in January 2009 [1]. The ratio of Bitcoin users to the total cryptocurrency users is 3:4. In the starting of 2017, the transaction count of alone Bitcoin exceeds all other cryptocurrencies transaction by 286,419 [2]. More than 40 countries (Germany, Croatia, Switzerland, Canada etc.) accepts Bitcoin as cryptocurrency [3]. To perform risk analysis before investing in bitcoin is recommended to the users because it is very difficult to predict its prices, which can be proven by the fact that in 2007, its price was 1,000 USD and in December 2017, it was 16,000 USD [1] [2]. Therefore, our research aim is to find the less time consuming and accurate model for the prediction of Bitcoin price from different "machine learning and deep learning algorithms" [4]. There are various factors like gold price, economic crisis, social media tweets that are continuously affecting the bitcoin price [5]. Factors open, close, high and low bitcoin price for a day are used in prediction work [6]. The dataset that contains per day interval rate, in US dollars from January 1, 2015 to July 8, 2020 was

Content from this work may be used under the terms of the Creative Commons Attribution 3.0 licence. Any further distribution of this work must maintain attribution to the author(s) and the title of the work, journal citation and DOI.
Published under licence by IOP Publishing Ltd

taken from Crypto Compare website and stored in JSON format in MongoDB (Big-Data Tool), and after that, some different regression and deep learning models are implemented. We have also developed a bitcoin price alert system using IOT, so that when the price of the bitcoin reaches the threshold value (the price at which user wants to sell his/her bitcoin) then it will send an alert to user through different ways like by sending the text message, and mail to the user and will also activate the buzzer.
This research uses six libraries. We used MongoDB, which is a Big-Data Tool, as a No-SQLdatabase, because to have a greater accuracy of the designed models we need huge data points, and to handle such a big data we need a Big-Data Tool like MongoDB [7].

i). NumPy
This library provides various methods to perform different operations on large datasets, n-dimensional arrays and matrix. [8].

ii). Pandas
This high-level library provides various methods to perform different operations on data frames [9].

iii). Matplotlib
Alibrary that provides various methods to plot different types of graphs.

iv). SciKit-Learn
This library provides various algorithms for predictive data analysis [10].

v). Keras
This library provides functionalities for developing different deep learning models and if required we can also use Tensorflow, CNTK, and Theano libraries with keras for developing high-level neural network program [11].

vi). Bolt
This library provides an easy to use interface of Bolt Cloud API's. It is used to communicate between Bolt Cloud API to Wi-Fi Module (Buzzer) and also to send SMS alert to the user.

vii). MongoDB
Eliot and Dwight developed MongoDB in 2007. It is a cross-platform, open-source, schema-less, NoSQL database which stores the data in document oriented approach in theform of key-values pairs and because of this the retrieval of data is also fast [12].

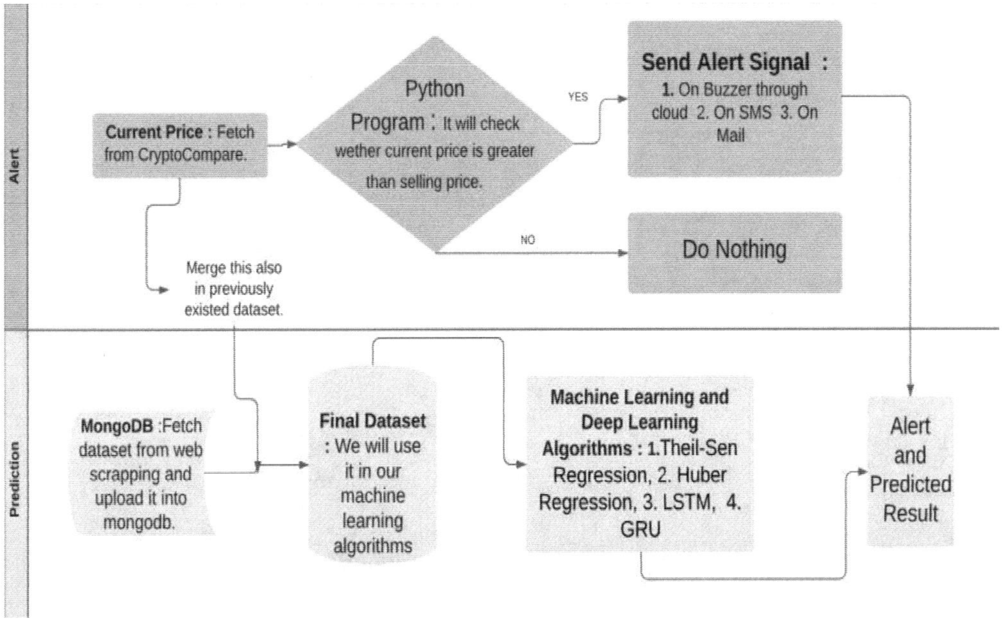

Figure 1. System Flow Diagram

2. Related Work

Phaladisailoed et al [1] applies and compares various regression models like Theil-Sen Regressor, Huber Regressor and different deep learning approaches like GRU, LSTM on the 1-minute interval trading exchange dataset from kaggle. The results show that GRU has highest R2 of 99.2% while Huber Regression has lowest execution time of 0.0002 seconds. Aggarwal et al [2] applied and compared different models of deep learning on different parameters that are affecting bitcoin price. The results show that LSTM has least RMSE value, sentimental analysis shows a positive correlation, and gold price analysis shows a negative correlation with the bitcoin price prediction.Rane PV et al [3] performs Bitcoin price prediction accuracy analysis by applying various models like ARIMA, LSTM, SVM, Latent Source model, Regression model, NARX model etc. The results show that NARX model has best accuracy. Greaves et al [13] uses Bitcoin transaction graph data for the prediction of Bitcoin prices. The exchange behaviour is not included in the transactions due to this the accuracy is 55%. McNally et al [14] has taken the data from Coin Desk and applies different models like ARIMA, LSTM and RNN. The result shows that highest achieved accuracy is 52.77% whereas the achieved RMSE value is 5.44%. Singh H et al [15] compares Support Vector regression (linear regression model) and KNN regression (polynomial regression model) based on root mean square error, and the result shows that KNN performs better than SVM.

3. Methodology

3.1. Data Collection

The 1-day interval trading dataset was collected from the crypto-compare website. The dataset contains 1501 data points from August 6, 2016 to September 14, 2020 in US Dollars.

3.2. Feature Selection

By calculating the correlation coefficient of all features and observing the pair-plot and heat-map correlation graph as shown in Fig 2, 3 and Table 1, we have selected the following features: close (closing value of bitcoin), open (opening value of bitcoin), high (highest value of bitcoin in a day), low (lowest value of bitcoin reached in a day). Next, we divide our dataset (1501 data-points) randomly in the ratio of 80:20 using train-test split. So, training dataset contains 1200 data points and the test dataset contains 301 data points.

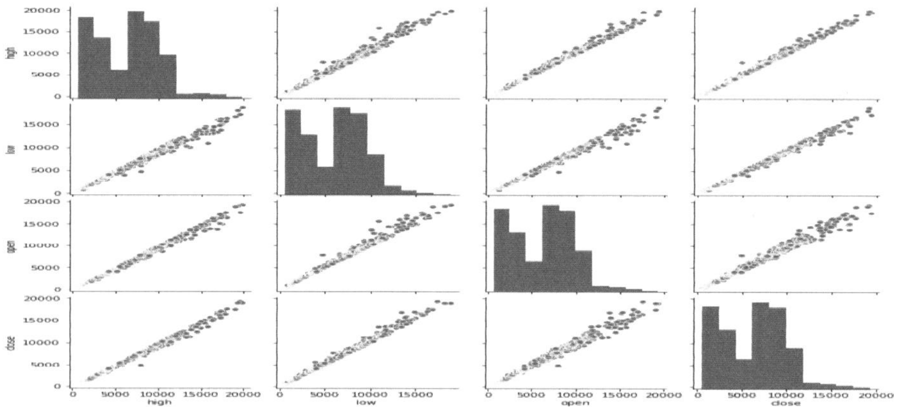

Figure 2. Correlation Pair-plot graph for each feature

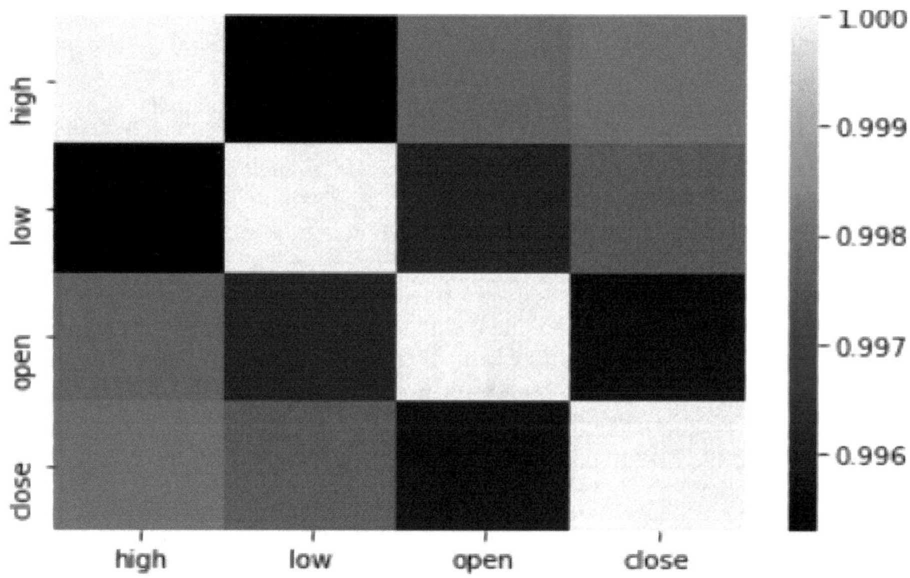

Figure 3. Heat-map plot of features

Table 1. Correlation coefficient matrix for each feature

	high	low	open	close
high	1.000000	0.995285	0.997839	0.998047
low	0.995285	1.000000	0.996269	0.997598
open	0.997839	0.996269	1.000000	0.995761
close	0.998047	0.997598	0.995761	1.000000

3.3. Data preparation

To adjust the data in the range of 0 to 1 we have used Min-Max Scaler [16] of scikit learn library and the equation it uses is:

$$Xnew = (Xi - \min(X)) / (\max(X) - \min(X))$$

Then time dimension was added to convert the 2-dimensional data into 3-dimensional data, to use it for "LSTM and GRU models" [3].

3.4. Modeling

As the Bitcoin prices are sequential in nature, so we have implemented and compared "machine learning models like Theil-Sen Regressor, Multivariate Linear Regression and Huber Regressor" [1] and for this purpose, scikit library was used, and deep learning models like "long-short term memory, gated recurrent unit" [7] and for this purpose we have used Keras library [1].

3.4.1. Multivariate Linear Regression

It is a parametric approach used to find how a response variable and several predictor variables are linearly related. Model estimates the coefficients of values. Exclusion of wrong values by various techniques helps to improve the efficiency of model [5]. Equation of multivariate linear regression is:

$Y = b0 + b1X1 + b2X2 + \dots + bnXn$

Y - response variable

X1, X2,Xn - predictor variables

b0 – y-axis intercept

b1, b2 – coefficients of predictor variables

n – no. of observations

3.4.2. Theil-Sen Regression

It is a non-parametric statistical method that tries to fit a best fit line in a 2-dimensional graph in such a way that the slope of the best fit line will be equal to the "median of the slopes of all lines possible through pairs of data points" [2]. It is strong against outliers because it can allow 29.3% outliers [17].

3.4.3. *Huber Regression*

It is a statistical regression technique that is robust to outliers. To distinguish between the outliers and inliers it uses a linear loss function rather than the least-square function. An inlier's weight is greater than outlier's weight [18].

3.4.4. *Long- Short term memory*

"It is a type of artificial recurrent neural network" [6] that has feedback connections. "Input gate, output gate, forget gate" [6] and a cell together makes a LSTM unit. LSTMs were "designed to overcome the vanishing gradient problem of traditional RNNs" [8]. The advantage of using a rolling window LSTM is that it recognises the sequential feature in the time series data that remains unnoticed by other algorithms. [4].

The equations of LSTM are:

$F_t = \sigma_g(W_f X_t + U_f H_{t-1} + B_f)$

$I_t = \sigma_g(W_i X_t + U_i H_{t-1} + B_i)$

$O_t = \sigma_g(W_o X_t + U_o H_{t-1} + B_o)$

$F_t = F_t O C_{t-1} + I_t O_{\sigma c}(WcXt + Uc\ H_{t-1} + B_c)$

$Ht = OO_t(C_t)$

where the values of C0 and h0 are equal to zero and the operator o is called Hadamard-product and t in the subscript denotes the time step.

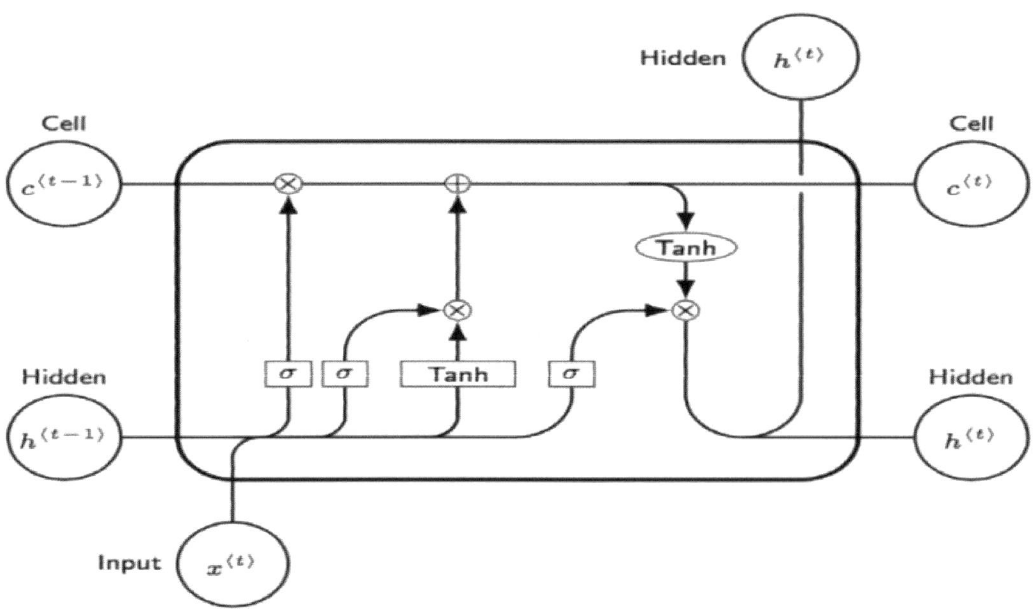

Figure 4. Structure of LSTM cell

3.4.5. Gated Recurrent Unit

Kyunghyun Cho in 2014 introduced Gated recurrent units (GRUs) in recurrent neural networks [20]. GRU are less complex than LSTM because they do not have an output gate but they contain a forget gate. The update vector and reset vector decides the flow of information. They can store the information for a long time, without washing it with the passage of time or removing the irrelevant part of it.

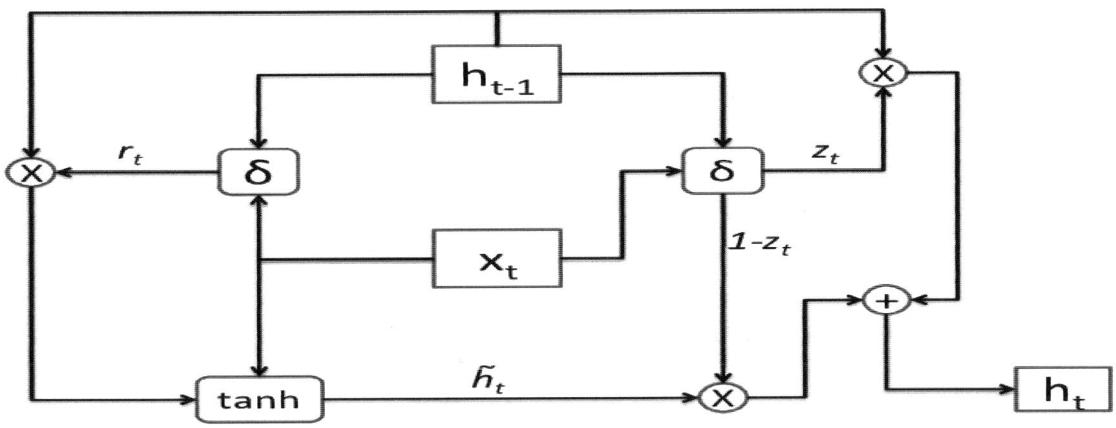

Figure 5. Structure of GRU cell

Equations of GRU:

Zt = σg(WzXt + Uz Ht-1 + Bz)

Rt = σg(WrXt + Ur Ht-1 + Br)

Ht = OHt-1Zt + (1-Zt)Oσh(WhXt + Uh (RtOHt-1) + Bh)

Variables:

Xt : input vector

Ht : output vector

Zt : update gate vector

Rt : reset gate vector

W,U and B parameter matrices and vector

E. IOT Implementation

Wifi Module

ESP-12E (ESP8266MOD)

3.4.6. ISM 2.4GHz

PA +25dBm : This is used as GPIO (General Purpose Input Output) device, mainly output device and uses UART(Universal Asynchronous Receiver Transmitter), SPI and I2C protocols. During this whole research we use only UART protocol for communication with buzzer attached with this Wifi module.

4. Results

Table 2 tells about the accuracy and execution time of all the implemented machine learning models whereas Table 3 tells about the MSE, R2 value and execution time of all the deep learning implemented models. The results show that among ml implemented models, Huber regression has the better accuracy and linear regression has best execution time, while among dl implemented models, GRU has better MSE and R2 value but execution time of LSTM is better than GRU.

Table 2. Accuracy and execution time of machine learning model

	Model_Applied	Accuracy	Execution Time
0	Linear_Regression	0.998647	0.011677
1	TheilSen_Regression	0.998507	1.762166
2	Huber_Regression	0.998668	0.078603

Table 3. MSE, R2 and execution time of deep learning models

	Model_Applied	MSE	R2	Execution Time
0	GRU	0.000085	0.997855	43.991490
1	LSTM	0.000087	0.997797	38.761662

5. Future prospects and Conclusion

The result of our research is, all models have same accuracy, but Linear Regression model has best execution time. However, there are several other factors like twitter sentiment analysis, gold price analysis, economic crisis, setting parameters, different policies and laws of different countries which can affect the results. Therefore, to obtain the best results always collect the updated data. In future, there is further scope in our research like "the accuracy of various models can be increased" [4], by applying new models, by using more features to predict the bitcoin price.

References

[1] Phaladisailoed T, Numnonda T. Machine learning models comparison for bitcoin price prediction. In2018 10th International Conference on Information Technology and Electrical Engineering (ICITEE) 2018 Jul 24 (pp. 506-511). IEEE.
[2] Aggarwal A, Gupta I, Garg N, Goel A. Deep Learning Approach to Determine the Impact of Socio Economic Factors on Bitcoin Price Prediction. In2019 Twelfth International Conference on Contemporary Computing (IC3) 2019 Aug 8 (pp. 1-5). IEEE.
[3] Rane PV, Dhage SN. Systematic erudition of bitcoin price prediction using machine learning techniques. In2019 5th International Conference on Advanced Computing & Communication Systems (ICACCS) 2019 Mar 15 (pp. 594-598). IEEE.
[4] Huisu J, Lee J, Ko H, Lee W. Predicting bitcoin prices by using rolling window LSTM model. In2018 Data Science in FinTech (DSF) 2018.
[5] Bisht A, Agarwal P. Analysis of Bitcoin using Linear Regression and Data Mining Techniques. International Journal of Advanced Research in Computer and Communication Engineering (IJARCCE), Department of Computer Science and Engineering, SRM University Chennai India. 2017 Nov;6(11).
[6] Mahrishi, M., Hiran, K. K., Meena, G., & Sharma, P. (2020). Machine Learning and Deep Learning in Real-Time Applications. IGI Global. http://doi:10.4018/978-1-7998-3095-5
[7] Rizwan M, Narejo S, Javed M. Bitcoin price prediction using Deep Learning Algorithm. In2019 13th International Conference on Mathematics, Actuarial Science, Computer Science and Statistics (MACS) 2019 Dec 14 (pp. 1-7). IEEE.
[8] Radityo A, Munajat Q, Budi I. Prediction of Bitcoin exchange rate to American dollar using artificial neural network methods. In2017 International Conference on Advanced Computer Science and Information Systems (ICACSIS) 2017 Oct 28 (pp. 433-438). IEEE.
[9] Jain A, Tripathi S, DharDwivedi H, Saxena P. Forecasting price of cryptocurrencies using tweets sentiment analysis. In2018 Eleventh International Conference on Contemporary Computing (IC3) 2018 Aug 2 (pp. 1-7). IEEE.
[10] Buitinck L, Louppe G, Blondel M, Pedregosa F, Mueller A, Grisel O, Niculae V, Prettenhofer P, Gramfort A, Grobler J, Layton R. API design for machine learning software: experiences from the scikit-learn project. arXiv preprint arXiv:1309.0238. 2013 Sep 1.
[11] François C. Keras: Deep learning library for theano and tensorflow. URL: https://keras. io/k. 2015.
[12] Parker Z, Poe S, Vrbsky SV. Comparing nosqlmongodb to ansql db. InProceedings of the 51st ACM Southeast Conference 2013 Apr 4 (pp. 1-6).
[13] Greaves A, Au B. Using the bitcoin transaction graph to predict the price of bitcoin. No Data. 2015 Dec 8.
[14] McNally S. Predicting the price of Bitcoin using Machine Learning (Doctoral dissertation, Dublin, National College of Ireland).
[15] Singh H, Agarwal P. Empirical analysis of bitcoin market volatility using supervised learning approach. In2018 Eleventh International Conference on Contemporary Computing (IC3) 2018 Aug 2 (pp. 1-5). IEEE.
[16] Patro S, Sahu KK. Normalization: A preprocessing stage. arXiv preprint arXiv:1503.06462. 2015 Mar 19.
[17] Dang X, Peng H, Wang X, Zhang H. Theil-sen estimators in a multiple linear regression model. Olemiss Edu. 2008.
[18] P. Huber. "Robust statistics." International Encyclopedia of Statistical Science. Springer Berlin Heidelberg, 2011, pp. 1248-1251.
[19] Hochreiter S, Schmidhuber J. Long short-term memory. Neural computation. 1997 Nov 15;9(8):1735-80.
[20] Cho K, Van Merriënboer B, Gulcehre C, Bahdanau D, Bougares F, Schwenk H, Bengio Y. Learning phrase representations using RNN encoder-decoder for statistical machine translation. arXiv preprint arXiv:1406.1078. 2014 Jun 3.

IOP Conf. Series: Materials Science and Engineering 1131 (2021) 012010 doi:10.1088/1757-899X/1131/1/012010

Comparative analysis of machine learning techniques based on chronic kidney disease dataset

Vineeta Gulati[1], Neeraj Raheja[2]

[1]Research Scholar, CSE Department, Maharishi arkandeshwar (Deemed to be University), Mullana 133207, India
[2]Associate professor, CSE Department, Maharishi Markandeshwar (Deemed to be University), Mullana 133207, India

vineetaarora04@gmail.com neeraj_raheja2003@mmumullana.org

Abstract. Kidney failure is one of the chronic diseases that becoming a very common health issue in the world. It is a state in which kidney is damaged and cannot clean blood as well as they should. Excess fluid and waste may cause more health issue in our body and takes long period to diagnose. Moreover, with no early symptoms, the disease is detected only at a later critical end stage. As this Chronic Disease is becoming threat in today's world, research is being done on a large scale to predict the presence of this chronic disease using machine learning. Machine learning is playing such a tremendous role in healthcare system like identify diseases and diagnoses, drug discovery and manufacturing, smart health record with the use of various techniques like Support Vector Machine, Decision Tree, Naïve Bayes, Random Forest etc. This paper comparatively analyzes the accuracy of pre-existing techniques for prediction of chronic kidney disease based on data from various research papers. Furthermore, this study also considers the different attributes from either already existing database or from real life database by using multiple techniques of machine learning. It is concluded that working with real life datasets with all possible attributes taken into consideration yields the accurate prediction for the presence of chronic kidney disease using machine learning.

1. Introduction

Kidney is very essential part of our body or we can say that we cannot survive without kidney because it filters waste and excess fluid from our body. But now a days kidney failure becoming a very common health issue in the world. It is a chronic disease means long lasting or takes long period to diagnose [1]. Chronic Kidney Disease is a state in which kidney is damaged and cannot clean blood as well as they should. Excess fluid and waste may cause more health issue in our body and this problem threatens all over the world [2]. As kidney disease progresses and irreversible destruction for our body, it may finally lead to kidney failure, which requires dialysis or a kidney transplant for survival In India 1 in 10 people estimated to be suffering from chronic kidney disease. Early analysis can lead to prevention of kidney failure. The best way to evaluate kidney function or guess kidney disease stages are to observe the Glomerular Filtration Rate (GFR) regularly [3]. GFR is calculated using age, gender, race and blood creatinine level.

Five stages of chronic kidney disease on basis of GFR:

Content from this work may be used under the terms of the Creative Commons Attribution 3.0 licence. Any further distribution of this work must maintain attribution to the author(s) and the title of the work, journal citation and DOI.
Published under licence by IOP Publishing Ltd

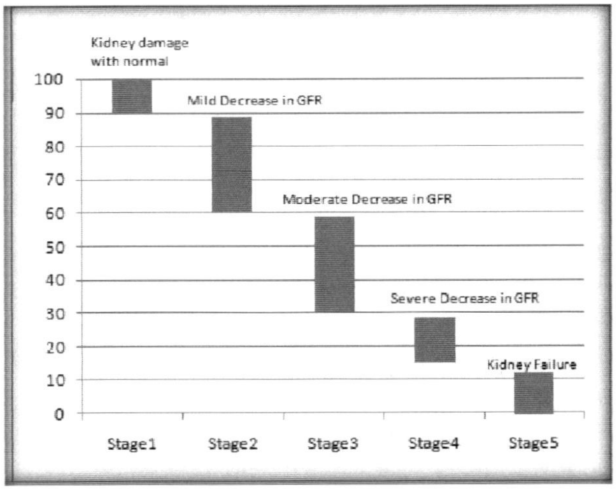

Figure 1. Stages of chronic kidney disease.

Kidney disease is usually asymptomatic early and can go undiagnosed until it develops. It is often referred as silent disease [4].

- **Stage 1**: Normal kidney function stage is when calculated GFR is greater than 90 ml(GFR>90ml).
- **Stage 2**: If approximate GFR is in between 60-80 ml then minor loss of kidney function started
- **Stage 3**: If calculated GFR is in between 30-59 ml then minor to harmful stage is to be considered.
- **Stage 4**: If approximate GFR is in between 15-29 ml then we can predict that the major loss of kidney function started.
- **Stage 5**: If calculated GFR less than 15 ml(GFR<15 ml) then this is the last stage of kidney failure which needs dialysis or transplant of kidney for survival.

Dialysis: When 90% of your kidney function fail then you require a artificial process to clean your blood from spare fluid, we have two process for this implementation. When an artificial kidney is used to fritter away fluid from your blood then this is called hemodialysis[5]. But when through pistula a tube pipe is fit in your arm or belly to purify your blood then this called peritoneal. Both the process is performed under the doctor supervision.

End-Stage Renal Disease (ESRD): When transplantation or dialysis is only option to treat complete and stable kidney failure.

Glomerular Filtration Rate (GFR): The rate at which the kidneys filter out waste and surplus fluid; calculated in millilitres per minute.

Proteinuria: When urine contains higher levels of an unusual protein called albumin.
Though, Kidney disease is asymptomatic in early stage but if we are able to diagnose this disease in early stage through accurate medical action then we can prevent chronic kidney[6]. If a professional system guess the kidney disease by examine the patient's symptoms like blood, urine test etc[7]. Then it gives doctors a proper time to heal or cure the patient disease on time. In current scenarios machine learning algorithms are playing important and accurate role in healthcare sector.

2. Machine learning

Machine Learning is outstanding within Artificial Intelligence area, which focuses more on to develop system which performs intelligently and also it has covered the largest real life impact for business

[8]. The main function of ML is to enable machine to work in self learning mode without being programmed explicitly, With the help of its algorithm implementation it make program able to learn, grow and change by themselves when exposed to new data. If the measurable performance of some task that is assigned by computer program is improve as it gain more and more experience then it called as Machine Learning or we can say that when machine take decisions and does prediction on data then also it is said to be the machine learning concept.
Classification, Regression and Clustering this type of problem are mainly solved through Machine Learning [9]. Which technique we are going to use is mainly depend upon the type and category of data we will provide to our model. The offered Techniques are Supervised, Unsupervised, Semi- Supervised and Reinforcement to implement machine learning algorithm.

2.1 Supervised Learning
It is the simplest and easiest type of learning method. When your dataset act as a teacher or you can say guide your model for training then this type of learning called supervised learning[10]. In this approach your model gets trained automatically and starts making prediction and take decisions.

2.2 Unsupervised learning
When your model try to find pattern from the given dataset or you can say that it learn through its observations and find the structure in the data then this type of learning is called unsupervised learning[11][12].

2.3 Semi supervised learning
From the name it is clear that it lies between supervised and unsupervised learning. Practically cost of label is very high in some situation so we need that much skilled person those can work with only few labels because majority case of labels are absent.

2.4 Reinforcement
This type of learning is completely different from supervised and unsupervised learning. Here in this learning we perform a feedback loop between an agent and an environment [13]. To make relation or to connect agent or environment we give set of actions. Video game is the best example of Reinforcement. Reinforcement goes through following steps:

- Agent observed the input state.
- Agent perform an action instructed by decision making function.
- Action performed and now agent receives reinforcement from the environment.
- This state-action information is stored for further use.

3. Machine learning in healthcare

Technology world's new inventions giving opportunity for people to develop a system which help the doctors to detect the chronic disease in early stage and cure the disease, Machine learning is playing so tremendous performance in healthcare system [5][14]. There are many application of machine learning in healthcare sector like identify diseases and diagnoses, drug discovery and manufacturing, medical imaging diagnoses, smart health record, better radiotherapy etc[8]. There are many chronic diseases in this world like cancer, lungs problem, heart disease, kidney failure etc. With no early symptoms and people comes to know only at the end stage, so here machine learning actually providing so many solution to early detect your problem and cure them.

3.1 Requirement of Decision Support in Healthcare
Many people die every year because of not able to detect disease early in the health care system. Health information technology construction suggested few strategies, like association, significant consumer selection of clinician society and IT adoption [15].

3.2 Decision Support

Health protection system in machine learning depends on the computer's large computing capacity and on doctor's logic competence [13]. Doctors and machine both search for pattern but doctors cannot estimate heartbeat of each patient as efficient machine can do. So, machine will do all these tasks and present the outcome for confirmation to doctor. *Decision support system in healthcare.* Decision support system help in early disease exploration by maintaining the awareness of health issue, background knowledge of individual patient to doctors [12]. It keeps track of insurance policy, refund detail, charge, account receivable, account payable. This system also helps in Identification of patient's situation and then doctor suggest when and how to use the drug to cure your disease.

4. Literature review

This table present the work done for prediction of chronic kidney disease using machine learning, here I, Study how dataset collection (Open Source or Real life Data) and their pre-processing affects the accuracy level of our study.

Table 1. Analysis of pre-existing prediction model for chronic kidney disease.

Author name	Study purpose and concept theoretical model	Repository used	Research findings
Ramya S and Radha N [2]	Presented different stages of kidney failure on basis of GFR and by using Random Forest, Radial Basis Function and Back propagation neural network classifier to identify which level of chronic disease is present in patient. Radial Basis Function gives maximum accuracy of 85.3%.	Data is collected through different medical reports of patients available in different laboratory in Coimbatore. 1000 records with 15 different attributes like age, gender, cretanine level, uric acid etc.	On basis of sensitivity, specificity, accuracy and kappa Radial Basis Function provide the accuracy of 85.3%.
Alassaf R A , Alsulaim K A, Alroomi, N Y, Alsharif N S, Aljubeir M F, Olatunji S O and Alturayeif N S [6]	The focus question is to predict CKD or not by two attribute selection process one is Correlation coefficient and another is recursive attribute elimination.	Data is Collected from King Fahd University hospital(KFUD) , 244 Records earlier with 491 features but after all refinement and cleaning it contains only 57 features.	ANN, SVM and Naïve Bayes provide 98% accuracy. **Limitation:** Only creatinine and urea correlation give highest accuracy, many attributes are ignored in this study to achieve the maximum accuracy.
Tekale S, Shingavi P,Wandhekar S and Chatorikar [7]	Presented a model for identify chronic kidney disease in patient by using pre existing dataset of patient and clean the data to achieve the better result, which is provided to our model for implementation process.	Datasets downloaded from UCI repository, 400 patient records with 25 aspect. Attributes used are red blood cell, Blood Pressure, white blood cell etc. Raw data needs to be clean so that properly use for our model like missing value removed by WEKA function "Replace Missing Value" with NA or by	Accuracy level given by Decision Tree is 91.75% and SVM is 96.7%. Calculation time is less. **Limitation:** Strength of the data is not higher because of size and missing aspect.

		mean value of that attribute. 14 attributes are used after pre-processing process with 300 patient record used for training of algorithms and 100 record are used for test purpose.	
Aljaaf A J, Al-Jumeily D, Haglan H M, Alloghani M, Baker T, Hussain A J and Mustafina J [8]	The purpose of study is develop a CKDPS system by using different approach like finds the correlation between input parameter for example cretanine and urea have strong relationship. On basis of urine test or blood test try to predict CKD by finding correlation between attributes so that to remove redundancy and noise in data and at last provide only correlated data to model for better result of prediction. Machine learning classifiers used in this study are regression tree, support vector machine, logistic regression and multi- layer perception neural network.	Dataset given by Apollo Hospital available online at UCI Machine learning Repository (400 Record of patient with 24 attributes). This study start with 24 attributes but after removing all missing values , noise , redundancy left only with 7 parameter with strongly correlated to each other given to the model for implementation.	Multilayer perception neural network give 99.5% accuracy with 98% sensitivity and 40 millisecond time consumption. **Limitation**: Value of most of the attributes is neglected to achieve the better accuracy but may some attributes affected by prescribed drug like blood pressure of patient affect and our model predict wrong prediction.
Revathy S, Bharathi B, Jeyanthi P and Ramesh M [13]	Examine the model for chronic disease prediction by using decision tree, random forest and support vector machine algorithms of machine learning. Adaboost and Logitboost two boosting techniques of data mining are used to enhance the performance of model. R programming tool are used for statistical analysis and confusion matrix with two dimension CKD or NONCKD is generated For clean data replace all the numerical value by mean or missing value by mode and then provide the dataset to model for training as well as testing purpose.	400 patient record with 25 aspect downloaded from UCI repository. 70% of dataset used for training of models and 30% used for testing processing.	Random forest provide the accuracy level of 99.16%. **Limitation:** Confusion matrix mean two or more classes for output prediction, in that case it is very difficult to predict that both the class equally involved or all the accuracy may belongs to only one class.
Feng B , Zhao Y Y, Wang J, Yu H, Potu S, Wang J and Guo Y[16]	Presented an independent way to predict CKD without GFR measurement, which basically overcome the incorrect and ambiguous issues related to GFR.	703 patient record, with all metabolite feature on basis of blood sample used from randomized hospitals.	With all metabolite feature RNN gives 95.6% accuracy SVM 95.4% Decision Tree – 95.4%.

	Random forest, Decision tree and support vector machine run parallel with randomized input values by 10 fold cross validation. A large 16282 metabolite feature dataset with clinical blood samples were used independently and after their prediction then only selected 69 metabolite feature dataset is used and conclude the difference. Further we apply unsupervised learning to validate that with selected metabolite is relate to critical factor of CKD.		With selected metabolite feature Models give maximum accuracy of 98%.
Anandajayam P, Aravindkumar S, Arun P, Ajith.A [17]	Purposed both a structural and un-structural real life data for early detection of chronic kidney disease. A big real life data plays a major role for better output Naïve bays, KNN, SVM and RNN algorithms are implemented on refined data.	Real life data of 43400 records collected from hospitals. A Digital format of patient's record like statistical data information is also used called electronic health record. 80% of dataset use for training and 20% use for testing.	RNN algorithm provide 97.62% of accuracy to predict CKD. **Limitation :** Dealing with both structural and un- structural combine form may be time consuming so for better result in future use only MRI scan or R-RAY.

5. Accuracy analysis of various techniques

The following study shows the accuracy comparison of various techniques, using different attributes from either pre- existing database or real life dataset to predict the presence of chronic kidney disease in an individual.

- Figure 2 shows the accuracy comparison of SVM and DT algorithms with 14 attributes of 400 patient records downloaded from UCI Repository[18]. SVM provides the best result with 96.7% accuracy but strength of the data is not higher because of size and missing aspect.

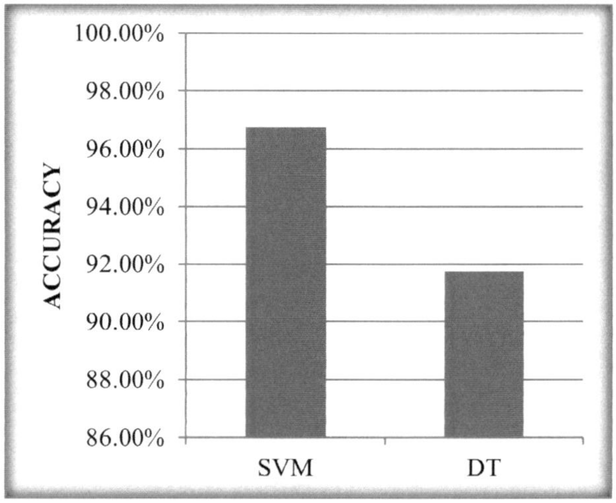

Figure 2. Accuracy comparison of SVM and DT techniques based on 14 attributes of UCI Repository

- Figure 3 shows the accuracy comparison of LOGR, SVM, MLP and RPART algorithms with 24 attributes of dataset downloaded from UCI Repository[19]. MLP gives the best result with 99.5% accuracy but finds only the correlation between input parameter to get the desired result for example cretanine and urea have strong relationship, Value of most of the attributes is neglected to achieve the better accuracy.

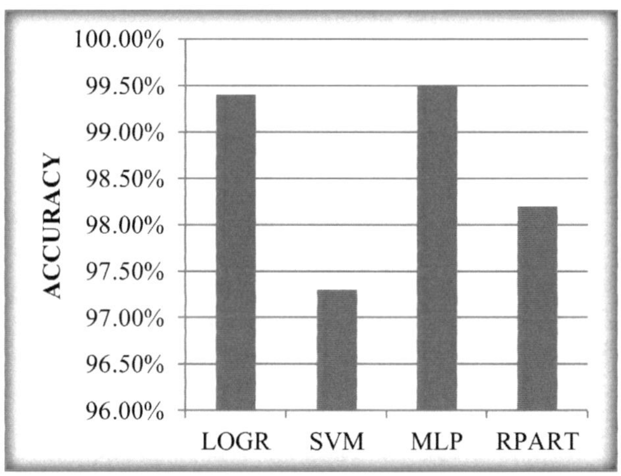

Figure 3. Accuracy comparison of SVM, LOGR, MLP and RPART techniques based on 24 attribute of UCI Repository dataset.

- Figure 4 shows the accuracy comparison of SVM, DT, RF algorithms with 25 attributes of dataset downloaded from UCI Repository. RF gives the best result with 99.16% accuracy[20].

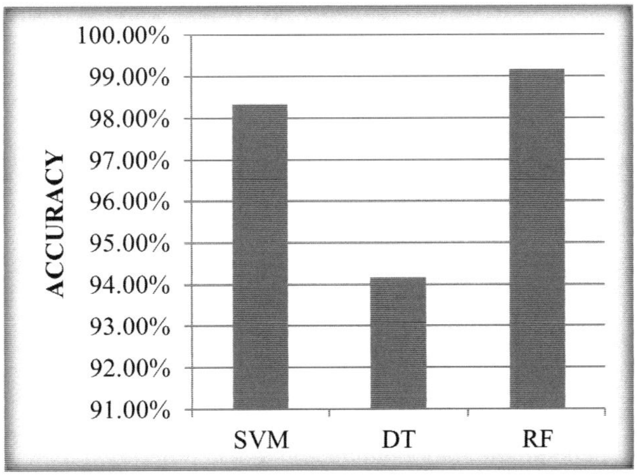

Figure 4. Accuracy comparison of SVM, DT and RF techniques based on 25 attributes of UCI Repository dataset.

- Figure 5 shows the accuracy comparison of DT, NB, KNN and RNN algorithms with 43400 records structural as well as un-structural collected from hospitals[21]. RNN gives the best result with 97.62% accuracy. This may be time consuming so for better result in future use only MRI scan or R-RAY.

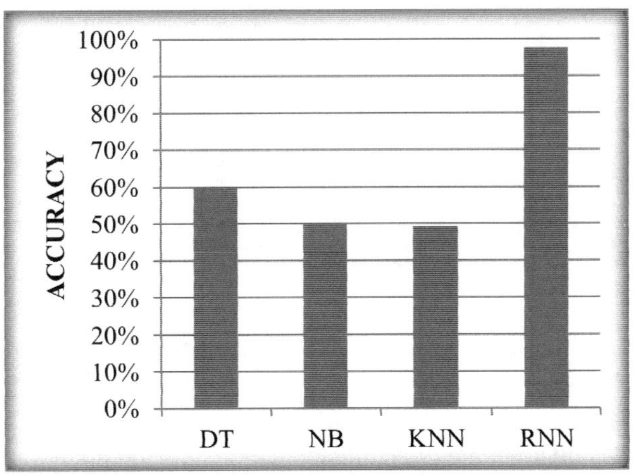

Figure 5. Accuracy comparison of NB, KNN, RNN and DT techniques based on structural as well as un-structural records.

- Figure 6 shows the accuracy comparison of SVM, DT and RF with 703 patient records, with all metabolite features on basis of blood sample used from randomized hospitals[16]. RNN gives the best result with 95.6% accuracy.

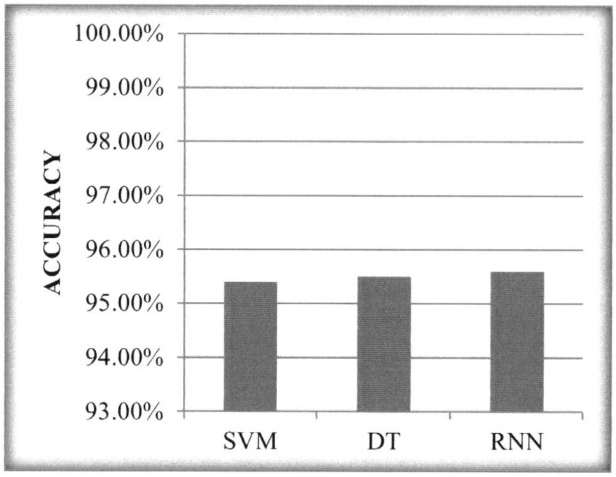

Figure 6. Accuracy comparison of SVM, DT and RNN with all metabolite features.

- Figure 7 Accuracy comparison of algorithms with selected 69 metabolite features[16] are shown below:

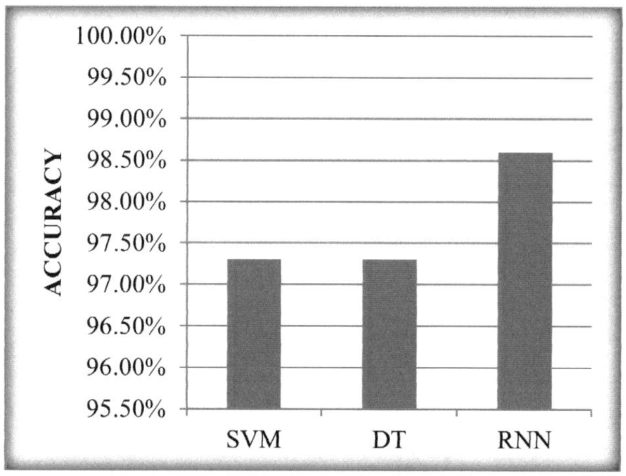

Figure 7. Accuracy comparison of SVM, DT and RNN with selected metabolite features.

6. Conclusion

This paper provides a Comparative Analysis of Machine Learning Techniques based on chronic kidney disease dataset with different attributes from either already existing database or from real life database. Numbers of machine learning algorithms are used like Support Vector Machine (SVM), Decision Tree (DT), Naïve Bayes (NB), Random Forest (RF), Recurrent Neural Network (RNF) etc. The study includes how dataset affect the accuracy level of an algorithm for instance, with only 14 attributes SVM results in the accuracy of 96.7%, with 24 attributes MLP results in the accuracy of 99.5%. However, the value of most of the attributes is neglected to achieve a higher accuracy[8]. In such as case, some attributes which are affected by factors such as prescribed drugs affecting blood pressure of the patient, we may have a high probability for a wrong prediction. Hence, to work with real life data with all the possible attributes for prediction of chronic kidney disease is more accurate, like as shown in the figure 6, with all metabolite features on basis of blood sample used from randomized hospitals, RNN gives the best result with 95.6% accuracy. Therefore, in the future working with real life datasets with all possible attributes taken into consideration yields the more accurate prediction for the presence of chronic kidney disease using machine learning.

7. References

[1] Sinha P and Sinha P 2015 Comparative study of chronic kidney disease prediction using KNN and SVM *International Journal of Engineering Research and Technology.* **4**(12), 608-12

[2] Ramya S and Radha N 2016 Diagnosis of chronic kidney disease using machine learning algorithms *International Journal of Innovative Research in Computer and Communication Engineering* **4**(1), 812-820

[3] Subasi A, Alickovic E and Kevric J 2017 Diagnosis of chronic kidney disease by using random forest *In CMBEBIH.* pp. 589-594 Springer, Singapore

[4] Shailaja K, Seetharamulu B and Jabbar M A 2018 March Machine Learning in Healthcare: A Review In *2018 Second International Conference on Electronics, Communication and Aerospace Technology (ICECA).* pp. 910-914 IEEE

[5] Ganiger S and Rajashekharaiah K M M 2018 December Chronic Diseases Diagnosis using Machine Learning *International Conference on Circuits and Systems in Digital Enterprise Technology (ICCSDET).* pp. 1-6 IEEE

[6] Alassaf R A , Alsulaim K A, Alroomi, N Y, Alsharif N S, Aljubeir M F, Olatunji S O and Alturayeif N S 2018 November Preemptive Diagnosis of Chronic Kidney Disease Using Machine Learning Techniques *International Conference on Innovations in Information Technology (IIT).* pp. 99-104 IEEE

[7] Tekale S, Shingavi P, Wandhekar S and Chatorikar A 2018 Prediction of chronic kidney disease using machine learning algorithm *Int. J. Adv. Res. Comput. Commun. Eng.* **7**(10), 92-96

[8] Aljaaf A J, Al-Jumeily D, Haglan H M, Alloghani M, Baker T, Hussain A J and Mustafina J 2018 July Early prediction of chronic kidney disease using machine learning supported by predictive analytics *IEEE Congress on Evolutionary Computation (CEC)* . pp. 1-9

[9] Mahrishi M., Morwal S. (2020) "Index Point Detection and Semantic Indexing of Videos—A Comparative Review". *In: Pant M., Kumar Sharma T., Arya R., Sahana B., Zolfagharinia H. (eds) Soft Computing: Theories and Applications. Advances in Intelligent Systems and Computing*, vol 1154. Springer, Singapore.

[10] Amirgaliyev Y, Shamiluulu S, and Serek A 2018 October Analysis of chronic kidney disease dataset by applying machine learning methods *12th International Conference on Application of Information and Communication Technologies (AICT).* pp. 1-4 IEEE

[11] Almasoud M and Ward T E 2019 Detection of chronic kidney disease using machine learning algorithms with least number of predictors *International Journal of Soft Computing and Its Applications* **10**(8)

[12] Ray S 2019 February A quick review of machine learning algorithms International conference on machine learning, big data, cloud and parallel computing (COMITCon). (pp. 35-39) IEEE

[13] Revathy S, Bharathi B, Jeyanthi P and Ramesh M 2019 October Chronic kidney disease prediction using machine learning models *International Journal of Engineering and Advanced Technology (IJEAT)* Volume-**9** Issue-1 *ISSN* 2249-8958

[14] Rabby, A. S. A., Mamata, R., Laboni, M. A., & Abujar, S. (2019, July). Machine Learning Applied to Kidney Disease Prediction: Comparison Study. In 2019 10th International Conference on Computing, Communication and Networking Technologies (ICCCNT) (pp. 1-7). IEEE.

[15] Dhillon A and Singh A 2019 Machine learning in healthcare data analysis: a survey *Journal of Biology and Today's World* **8**(6), 1-10

[16] Feng B , Zhao Y Y, Wang J, Yu H, Potu S, Wang J and Guo Y 2019 May The application of machine learning algorithms to diagnose CKD stages and identify critical metabolites

features *International Work-Conference on Bioinformatics and Biomedical Engineering* (pp. 72-83) Springer Cham
[17] De S and Chakraborty B 2019 September Development of chronic kidney disease prediction system (CKDPS) using machine learning technique *In International Conference on Intelligent Data Communication Technologies and Internet of Things* (pp. 155-164) Springer Cham
[18] Veerasekaran K and Sudhakar P 2019 September Machine learning based predictive models and its performance evaluation on healthcare systems *International Conference on Intelligent Data Communication Technologies and Internet of Things* (pp. 677-684) Springer Cham
[19] Almasoud M and Ward T E 2019 Detection of chronic kidney disease using machine learning algorithms with least number of predictors *International Journal of Soft Computing and Its Applications*. **10**(8)
[20] Xiao J, Ding R, Xu X, Guan H, Feng X, Sun T, and Ye Z 2019 Comparison and development of machine learning tools in the prediction of chronic kidney disease *Progression Journal of translational medicine*. **17**(1), 119
[21] Anandajayam P, Aravindkumar S, Arun P, Ajith A 2019 Prediction of Chronic Disease by Machine Learning *Proceeding of International Conference on Systems Computation Automation and Networking*.

IOP Conf. Series: Materials Science and Engineering 1131 (2021) 012011 doi:10.1088/1757-899X/1131/1/012011

A Pun Identification Framework for Retrieving Equivocation Terms based on HLSTM Learning Model

Shashi Shekhar[1], Rishabh Sharma[2]

Department of Computer Engineering & Applications,
GLA University, Mathura, India.

[1]shashi.shekhar@gla.ac.in,[2]rishabh.sharma_mtcs19@gla.ac.in

Abstract. The artificial intelligence system is made linguistically intelligent through HLSTM model to identify pun expressions from code mixed text. The text available on social media domain is written in mixed script format and from these content puns word identification is a challenging task in this scenario. The retrieval of pun and its corresponding equivocation terms is very hard to retrieve from the transliterated text. The pun retrieval and its equivocation representation are widely used to present the opinions over the network applications. The work described in the paper gives the comparative view of different learning models and techniques applied in the area of transliteration for pun word retrieval. The rule framed approach is presented which accepts the roman form text as input and as per the defined rules the system is developed to give the equivocation words available in the sentence. The evaluation measures used here to validate the hypothesis is based on statistical measures along with HLSTM learning model. Further the result is validated using the voting technique that can choose appropriate equivocation label which are not identifies by the learning model. The use of voting technique here is to provide an extra edge when the proposed approaches suggest incorrect tag against the pun word. The voting approach enhances the overall result accuracy with high precision value.

1. Introduction

Equivocal expressions now days are regularly used in the social media networks. Equivocal information processing is the challenging task in the transliterated domain. The equivocation helps in answering the ambiguous category of expressions by identifying the words based on its context meaning. Also it helps in illustrating the things in the terms of ambiguity order. Many researches has been undertaken to produce equivocal annotation of text in several languages. But very few works has been done in Indian language transliteration. The paper explores this area in retrieving ambiguity based information in transliterated Roman Hindi domain. Here the classification related to humor expressions in Roman Hindi is analyzed. The paper presents a voting based approach based on crafted rules for classifying equivocal humor data in internet domain.

The rules are modeled subject to regular expression matching. When the matching is done, it checks for ambiguous data and the words are tagged as equivocal expression. The labeling parameters consisting of equivocal dimension are checked against the rule and label that word as equivocal expression. Here in this work, we first analyze the structures of words in sentence to identify the ambiguity in code-mixed text and classify them as ambiguous or non-ambiguous. The language identification task is used here as we are concern only about the transliterated Hindi words that are checked in context to the entire sentence for identifying the equivocal words. For context identification technique of H-LSTM based framework has been designed using CBOW technique that

Content from this work may be used under the terms of the Creative Commons Attribution 3.0 licence. Any further distribution of this work must maintain attribution to the author(s) and the title of the work, journal citation and DOI.
Published under licence by IOP Publishing Ltd

targets majority of the ambiguous words. Language and word detection in the user-generated text is a very tedious task, where the language is un-known. Now a day, it is the difficult task where the text is available in the format of code-mixed sentence. This type of code-mixed data is common in the social media sites. The main challenge here is due to availability of many transliteration variants for a given word. Lastly we test our approach on a dataset of Amul advertisements in India [1] and the proposed framework is able to recover equivocal words.

The available detection systems are not equipped to deal with equivocal data. This paper describes the use of equivocal data to identify the language as well as the dimensions of the humor context in which it has been used in the expression. This identification is necessary for the languages which are linguistically much related with each other. To separate words that are syntactically similar in both languages, a special technique is required. The natural language is one of the medium for communication in India. The processing of this by the machine requires specialized skill to extract meaningful information based on humor dimension. It is an emerging area of research for extracting intent of the user for using ambiguous humor expression for expressing opinions. With the huge use of social media platforms for information exchange, it is likely to have natural language data that needs to be processed by the machine to get information. These platforms are widely used by Indians to discuss any issue especially using their own native languages. Previously we were using mainly English language for such communication but in present scenario peoples are using mixed script contents for information exchange. Now a day's in Indian scenario, people are mixing more than one language for expressions to be posted on social media. These scenarios are leading to the field of code-mixing. To better understand the scenario of code-mixing an example has been illustrated from the advertisement of Amul, which describes the exploration of equivocal expression in present time. Transliterated Hindi-English mixed text is described in the below sentence:

Sentence 1:	Namaske	President	Russia
	H/EQ	E	NE

Here, words in English as E, Roman Hindi are labeled as H, Named entity as NE and ambiguous equivocal as EQ. The ambiguous equivocal words describe the ambiguity expression in roman Hindi in sentence. In sentence 1 the word Namaske is labeled as EQ, it illustrates that the word denotes equivocal expression. The proposal describes an architecture that represents context level information for presenting the equivocal tag associated with context dimension words used in the sentence, which are marked as the Hindi word.

The rest part of this paper is structured as follows: section 2 illustrates the state of art in equivocal retrieval. The methodology description is available in section 3. The description of dataset and its corresponding evaluation is being contained in section. The section 5 provides the summarization and conclusion with future path of work.

2. Related work

In this section provides the literature survey in recent techniques regarding temporal information in transliterated domain.

Code-mixing is an emerging area of research in the area of language classification. Identifying the language is the major task for any linguistic processing applications. Presently several type of research is going on in the field of code-mixing. The proposal of King etal.[2] utilizes supervised mechanism for language identification. The paper [3] implements CRF model for identifying the language. The proposal of given in [4] uses logistic regression, in code-switching environment. Das et.al. [5] proposed the use of dictionary along with the concept of edit distance to find word origin in regard to word context.

The task conducted on Mixed Script Information Retrieval (MSIR), where language identification for Indian languages combined with other languages have been scheduled [6] focusing on the use of transliteration. The task of MSIR was evaluated using SVM attaining an accuracy of greater than 75% [7]. The proposal of [8] uses supervised learning for English-Hindi word identification. The use of

Naive Bayes classifier [9] was proposed for Hindi-English data. The paper [10] proposed embedding technique as a feature for entity extraction.

A mixed script based language identification task was conducted for Indian Languages [11]. Here the use of machine learning techniques using SVM classifier [12] was proposed. The technique of classification and its related machine learning techniques for English-Hindi [10][11] languages were taken care. This task gives the opportunity for the emerging researched to enhance their learning and understanding the domain area covered under transliteration field [13]. Various emotion identification models have been described based on learning-approach [13] for language mining [14].

The work consisting of ambiguity removal in code mixed text needs to be handled [15] with the help of learning models mostly in native language domain for finding effective context meaning. The following section of proposed methodology tries to model the ambiguity problem available in code mixed data using embedding technique [16]. The embedding model is more concerned for those words which are commonly used in both the languages. As it the most common research issue [19] in multilingual dataset [17] used in case of NER [18] extraction in transliterated domain.

The paper [14] suggests the work related to NER in multilingual environment based on learning model and correspondingly provides a research issue in the field of retrieving phrases having equivocal opportunity in transliteration.

3. Proposed Framework

The proposed framework is inspired by the latest work [20][21] undertaken in the field of language pairs that have different lexicons representing different context meanings when combined with other words. The research is underway based on the neural learning architecture to understand equivocation in finding humor with the help of pre-trained embedding technique for building the Recurrent Neural Network based transliteration model. The proposal presented in this paper is based on related research findings in the area of code-mixing. The intricacy to detect the language of the temporal expression words in code mixed data is presented in this work. The data which is present in the code-mixed format includes more than one script. Due to this mixing, complication in processing is bound to arise. Language classification with accuracy is the foremost problem identified in this case. The problem of identifying language in these domains is more complex as the text contents are written in different languages and it is difficult to identify humor equivocation information in such cases. The following section describes the equivocation architecture based on HLSTM model for extracting humor sense from the data.

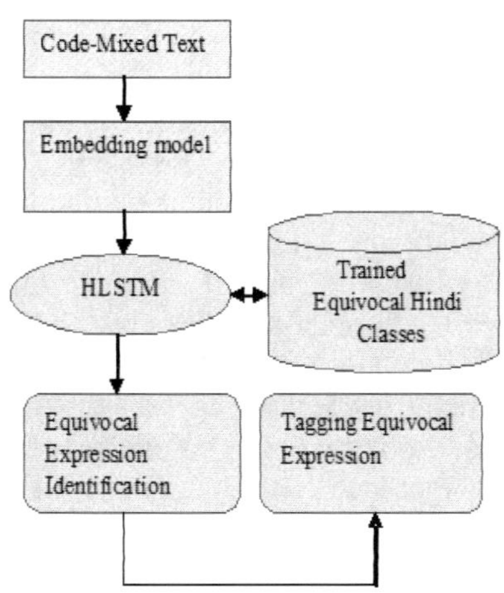

Figure 1. HLSTM Framework.

The figure 1 describes the proposed HLSTM system for equivocation data retrieval from code-mixed domain. The proposed methodology is trained at the word level using the HLSTM on the basis of various features selected for detecting the ambiguous equivocation words in the code-mixed data. The system takes code mixed input. The data is first tokenized as embedding process is based on word embedding and character embedding. Many probable equivocation classes exist for roman Hindi words. Character embedding is done for roman Hindi words as per the defined classes presented in table 1 to find equivocation expression. The token matching is done on the basis of parts of speech and words available in the input text for predicting equivocation expression based on hierarchical LSTM model. The features which can be used in the roman Hindi words using equivocation classes are given as the training set.

To understand this concept, Table 1 provides the in detail considering the equivocation classes to which the input word belongs to. Here backward and forward, LSTMs are used in the embedding layers. Finally in output layer softmax function is applied on the character vectors for labeling the token based on equivocation humor expression words. The model Hierarchical LSTM considers the neighboring words to the pivot word for suggesting the tag through context analysis features. The embedding technique used by the learning model which takes word and character embedding features for equivocation analysis.

Table 1. Equivocation expression classes.

Class	Examples
Puns in Headlines and Advertising	Happy Twenty Dineteen (Twenty)
Pun Time expression	3 gande (3 ghante/ Hours)
Puns in Quotes	Namaske(Namaste) Trump
Pun- Past, present and future expression	Pighle din(last day), is din(present day), pagle saal(next year)

The objective behind finding equivocation words available in mixed script text is summarized below:

- Ambiguous words retrieval in mixed text.
- Labeling the ambiguous words against the defined equivocal classes.
- Intent retrieval for using equivocal words.

3.1. Context retrieval equivocation expression

The consideration of the terms that exist prior to pivot term and next to pivot word forming the things as (i+1) term and (j+1) term, are used as word features for context finding. The Amul data [22] containing monolingual format sentences are used to frame the linguistic learning model for context tracing. The constraint applied for context tracing is that the equivocal words must be accompanied by the terms available in the left and right of that word should belong to other languages. The concept of intersection needs to be computed to find whether the word has been used in Hindi or English context. This modeling helps in finding the similarity which further helps in finding appropriate pun words.

The evaluation measure for context identification for finding the ambiguity is measured with the left and right context in regard to the used pivot word. The evaluation explores the base of the data discussed in for starting the self-learning approach. The condition used here in this case is that the left and right words to the pivot word that belong to two languages. The set theory intersection concept is applied for tagging. The context word is retrieved on the basis of WX notation. Thus considering this scenario the roman Hindi words describing different context has been pointed out in figure 2.

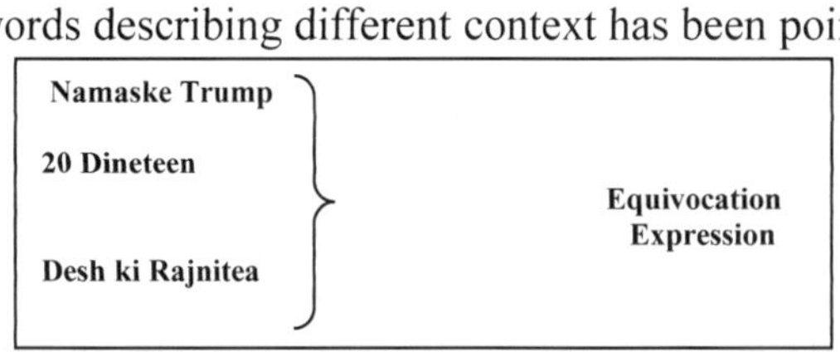

Figure 2. Equivocation context lookup.

3.2. Embedding model

The next step is to process input data against the embedding model. Word embedding is the weighted vectors of terms. The words can be represented in different dimensions and every term contains different weights in context to different dimensions. The meaning of word used as equivocation can easily be understood by the technique of embedding .This embedding technique of CBOW and Skip-gram technique will help to understand the context meaning of the word in connection with other word. Thus this technique helps in identifying the pun word used in the data set which has been used for evaluating the framework. CBOW technique is illustrated in figure 3.

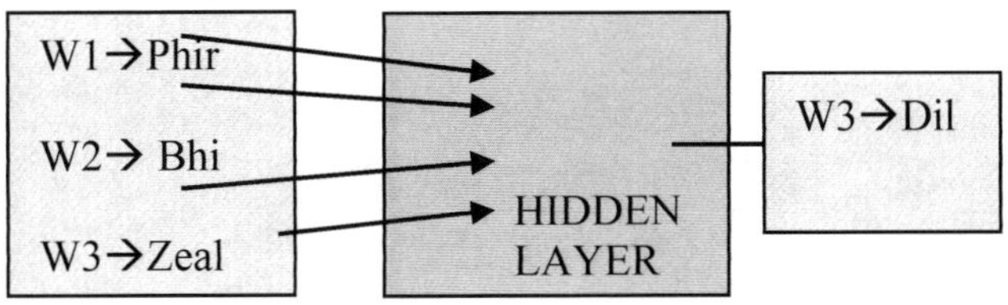

Figure 3. Embedding Technique.

3.3. Experiment Evaluations

This section describes the evaluation scheme undertaken for the proposed model depicted in figure 1. The result description is presented by illustrating the use of dataset and its inference in this section. The dataset of code-mixed data used here is taken from the work of ICON-2016 and Amul advertisements [22]. It is Hindi-English text containing data of three social media texts along with advertisements hits of Amul containing headlines of advertisements.. The data description is illustrated in table 2. The data of these media texts has been labeled for HLSTM learning on four dimensions considering equivocation expression as base for classification. The four labeling parameters are H-EQ for Hindi words, E-EQ for English words, N-EQ for not recognized words and O for words belonging to other than Hindi and English. The labeling parameters and its corresponding description along with percentage are depicted in table 3.

Figure 4 provides the result analysis obtained for labeling accuracy on table 3 parameters. The four labels depicted in table 3 are evaluated on the data available in table 2. The labeling accuracy as per the f-score obtained is higher in case of Hindi words as compared to available English words. The figure 3 depicts the F-score obtained on the dataset.

Table 2. Dataset description.

Code-mixed text	No. Of words
ICON-2016	2823
Amul Advertisements	1123
MSIR-2016	5867

Table 3. Labeling parameters

Tag	Description	Hindi-English (Mixing)%
H-EQ	Equivocation terms (Hindi)	58.21
EQ	Equivocation terms (English)	21.05
N-EQ	Equivocation words not recognised	13.8
O	Other language word	6.94

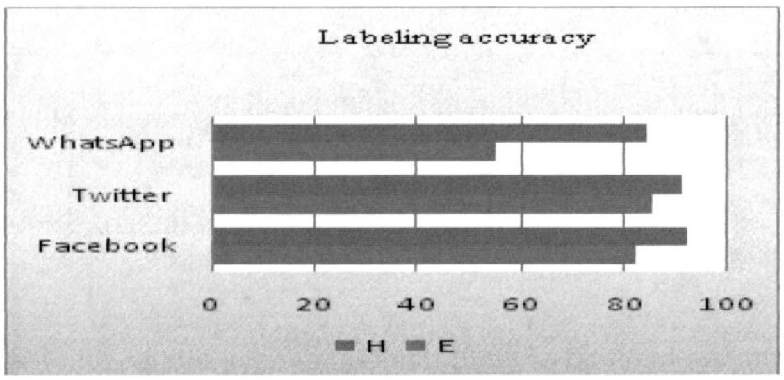

Figure 4. Labeling accuracy.

The Figure 5 provides a result description of the model in regard to embedding technique used for context lookup. The figure describes the accuracy percentage of embedding model trained for character and words. The proposed HLSTM gives a clear separation for the different equivocation classes parameters, depicted in table 3. The result is compared with five different standard classification mechanisms based on equivocal pun expression words. The accuracy of developed HLSTM provides enhanced accuracy when compared with standard measures.

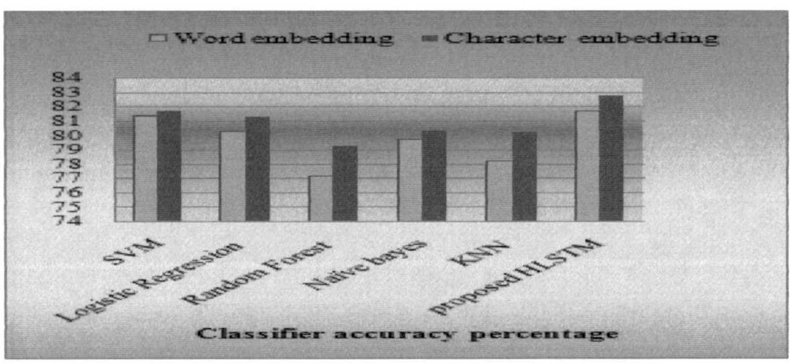

Figure 5. Classifier accuracy comparison.

The Figure 6 pointed below provides the detailing of occurrences of context words which has been selected for evaluation. These words are categorized as pun expression terms belonging to the defined data sample. These words exhibit equivocation features and are highly used for expressing opinions. Their contextual meaning can be different when correlated against the other terms available in the sentence.

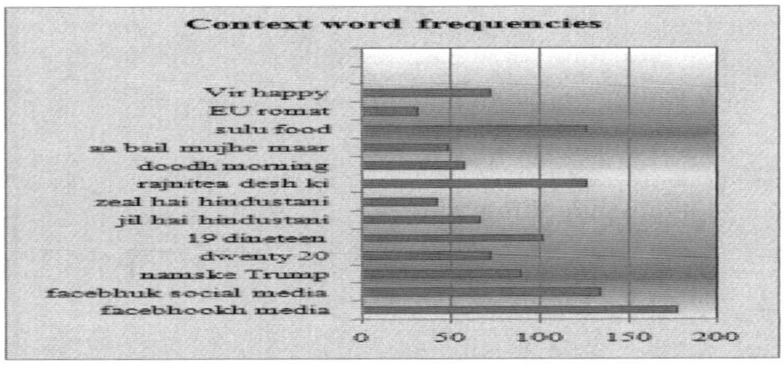

Figure 6. Equivocal expression words and their frequency.

The figure 7 depicts the framework accuracy when compared against the standard models like CRF, DT, SVM and rule based. The developed voting technique provides greater accuracy when compared with the CRF, DT, SVM standard models.

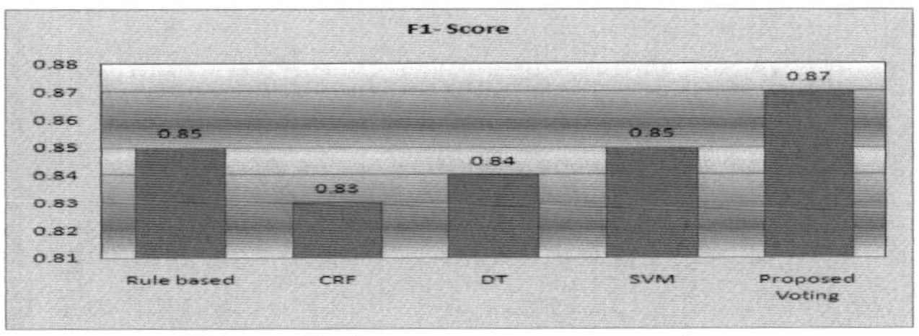

Figure 7. Accuracy representation.

4. Conclusion and Future Work

The paper shows that equivocal expression retrieval is one of the prominent areas in information retrieval, where one can understand the context by identifying pun expressions. The learning strategy based on predefined equivocal classes improves the labeling performance. This is one of the issues in language identification where equivocal expression or words need to be identified correctly in multi lingual environment. The multiple language use in code switching and code mixing environment is based on certain defined parameters like source of data, unstructured nature of data, switching and mixing percentages along with semantic relationship among the languages used for expression. We conclude that the equivocal expression words are often used on social media and advertisements according to the experiments conducted. It can be an interesting domain to investigate the patterns of words used to exhibit multiple contexts. The words which are used in Hindi as well as in English for expressions are needed to be examined for extracting equivocal dimension information. The paper provides state of art approaches for equivocal expression identification in Hindi. The paper illustrates different evaluation mechanism and comparison of proposed approach with standard approaches. It is being observed from the results that the proposed voting scheme gives better result in terms of F1 score. Our experiments were mainly on two language pairs based on bilingual learning approach. A HLSTM based learning approach has been proposed for classifying equivocal information in code-mixed text which gives better results. The system can be enhanced to learn other patterns in data like hate or satire detection in social media text.

References

[1] Mamidi, Radhika 2018 *Context and Humor: Understanding Amul advertisements of India* (*arXiv preprint arXiv:1804.05398*).

[2] King, Ben, and Steven Abney 2013 *Proceedings of the 2013 Conference of the North American Chapter of the Association for Computational Linguistics: Human Language Technologies* pp.1110-1119.

[3] Nguyen, Dong, and A. Seza Doğruöz 2013 Word level language identification in online

multilingual communication. *Proceedings of the 2013 conference on empirical methods in natural language processing* p 857.

[4] Vyas, Yogarshi, et al.2014 *Proceedings of the 2014 Conference on Empirical Methods in Natural Language Processing (EMNLP)* p 974.

[5] Das, Amitava, and Björn Gambäck 2014 Identifying languages at the word level in code-mixed indian social media text p 169.

[6] Sequiera, Royal, et al. 2015 Overview of FIRE-2015 Shared Task on Mixed Script Information Retrieval *FIRE Workshops*. **1587** p 19.

[7] Rao, Pattabhi RK, and Sobha Lalitha Devi 2016 CMEE-IL: Code Mix Entity Extraction in Indian Languages from Social Media Text@ FIRE 2016-An Overview *FIRE (Working Notes)* p 289.

[8] Shekhar, Shashi, Dilip Kumar Sharma, and MM Sufyan Beg 2018 Hindi roman linguistic framework for retrieving transliteration variants using bootstrapping *Procedia Computer Science* **125** p 59.

[9] Ethiraj, Rampreeth, et al. 2015 NELIS-Named Entity and Language Identification System: Shared Task System Description *FIRE Workshops* p 43.

[10] Alekseev, Anton, and Sergey Nikolenko 2017 Word embeddings for user profiling in online social networks. *Computación y Sistemas* **21** p 203.

[11] Mandal, Soumil, and Anil Kumar Singh 2018 Language Identification in Code-Mixed Data using Multichannel Neural Networks and Context Capture *arXiv preprint arXiv:1808.07118.*

[12] Aina, Laura, Kristina Gulordava, and Gemma Boleda 2019 Putting words in context: LSTM language models and lexical ambiguity *arXiv preprint arXiv:1906.05149.*

[13] Bhattacharya, Paheli, Pawan Goyal, and Sudeshna Sarkar 2018 Using Communities of Words Derived from Multilingual Word Vectors for Cross-Language Information Retrieval in Indian Languages *ACM Transactions on Asian and Low-Resource Language Information Processing (TALLIP)* **18** p 1.

[14] Shekhar, S., et al. 2020 Artificial Immune Systems-Based Classification Model for Code-Mixed Social Media Data *IRBM.*

[15] Murthy, Rudra, Mitesh M. Khapra, and Pushpak Bhattacharyya 2018 *ACM Transactions on Asian and Low-Resource Language Information Processing (TALLIP)* **18** p 1.

[16] Shekhar, Shashi, Dilip Kumar Sharma, and MM Sufyan Beg 2017 Linguistic structural framework for encoding transliteration variants for word origin detection using bilingual lexicon *2017 International Conference on Multimedia, Signal Processing and Communication Technologies (IMPACT)* p 156.

[17] Mahrishi, M., Hiran, K. K., Meena, G., & Sharma, P. (2020). Machine Learning and Deep Learning in Real-Time Applications. IGI Global.

[18] Shekhar, Shashi, Dilip Kumar Sharma, and M. M. Sufyan Beg 2019 An effective cybernated word embedding system for analysis and language identification in code-mixed social media text *International Journal of Knowledge-based and Intelligent Engineering Systems* **23** p 167.

[19] Shekhar, Shashi, Dilip Kumar Sharma, and MM Sufyan Beg 2020 "Computational linguistic retrieval framework using negative bootstrapping for retrieving transliteration variants." *International Journal of Computational Vision and Robotics* **10** p 79.

[20] Ramrakhiyani, Nitin, and Prasenjit Majumder 2015 Approaches to temporal expression recognition in Hindi *ACM Transactions on Asian and Low-Resource Language Information Processing (TALLIP)* **14** p 1.

[21] Shekhar, Shashi, Dilip Kumar Sharma, and MM Sufyan Beg 2020 Language identification framework in code-mixed social media text based on quantum LSTM—the word belongs to which language? *Modern Physics Letters B* Vol 34.

[22] Amul, https://www.amul.com/m/amul-hits. Accesssed, Dec 2020.

Performance Study of Various Machine Learning Classifiers for Arc Fault Detection in AC Microgrid

S Ramana Kumar Joga[1], Pampa Sinha[2], Manoj Kumar Maharana[3]

Research Scholar[1], School of Electrical Engineering, KIIT Deemed to be University, Bhubaneswar, India

E-mail: sanset567@gmail.com[1]

Abstract. Fault is the abnormal condition in power system, which must be detected as early as possible. It is very important to detect the fault as quick as possible to reduce the effects of fault like equipment damage, property loss and human loss. Arc faults have high power discharge property between two conductors; this property causes damage to the conductors which leads to electric fire between the conductors. It is very necessary to detect these faults immediately to avoid fire accidents. There are various methods to detect these arc faults in microgrid. In this paper voltage and current signals are measured through instrumental transformers and voltage signal is decomposed by the discrete wavelet transform signal processing technique. The decomposed signals are further processed in various machines learning classifier's for detecting the arc fault. The Proposed methodology studies the performance of various machine learning classifiers to detect arc fault in microgrid and it is carried out in MATLAB/Simulink Software.

Keywords: Arc Fault, Machine Learning, Wavelet, Microgrid, Fault Detection, Discrete Wavelet Transform, PQ disturbances.

1. Introduction

In Microgrid, sometimes abnormal condition changes the electrical flow in the System and causes change in the voltage and current levels. In other words, the sudden change in the direction of flow of current in electrical circuit is known as fault. There are many types of faults that occur in microgrid, they are short circuit faults, arc faults and open circuit faults and it is more discussed in [1]. The causes of fault occur in the power system due to failure of insulation, by human errors and due to bad weather conditions. The effects of electrical fault are that huge amount of current will flow through the equipment which produces more amount of heat and it damages the electrical equipment.Arc fault is one of the dangerous electrical hazards in power system. Arc fault is one of the leading causes of fires. Arc fault is virtually impossible to detect. Arc fault can't be detected by common protected device. The causes of arc fault occur in the power system due to damage, faulty insulation and loose in the terminal condition. Microgrid is a sub-set of smart grid. Microgrid provides small-scale power supply. Microgrid is designed to provide power for a small community. It gives the authority to local power generation for local loads. In Microgrid, there are two operating modes are grid connected mode and island mode. Transmission losses are reduced in power system through microgrid. Microgrid provides high quality supply. The importance of fault detection is that it can help to avoid abnormal conditions. In a Microgrid, many small circuits are connected to avoid the damage of interconnected active circuit. Through fault detection damage of equipment is avoided and small circuit which are interconnected in

Content from this work may be used under the terms of the Creative Commons Attribution 3.0 licence. Any further distribution of this work must maintain attribution to the author(s) and the title of the work, journal citation and DOI.
Published under licence by IOP Publishing Ltd

IOP Conf. Series: Materials Science and Engineering 1131 (2021) 012012 doi:10.1088/1757-899X/1131/1/012012

the Microgrid. By the help of fault detection electrical fires are avoided, protect from injuries and save time. There are many methods to detect fault in microgrid and they are discussed by Yashasvi Bansal Et.al in [2]. James J.Q et.al discussed the intelligent scheme of fault detection based on wavelet transform and deep neural network and it is furthermore discussed in [3]. Machine learning based fault detection in microgrid is discussed in [4].Meenakshi Mohan et.al discussed the digital protection of microgrid based on wavelet transform and it is discussed in [5]. Jayamaha et.al discussed the wavelet artificial neural network based fault detection in DC microgrid , and it is discussed in [6].Online fault detection using discrete wavelet transform and probabilistic neural network is discussed in [7]. In this Paper, a discrete wavelet transform based signal decomposition technique is used to detect the arc fault in microgrid through various machine learning classifiers. It provides the detail study on the performance of various machine learning classifiers in performing fault detection in microgrid task.

2. Proposed Methodology

The fault detection in Microgrid basically involves three steps, they are

a. Voltage and Current Signal Measurement through Voltage and Current Sensors
b. Decomposition of Voltage and Current Signals through Signal Processing Technique.
c. Training of data through machine learning classifiers.

The basic fault detection block diagram is shown in Fig. 1.

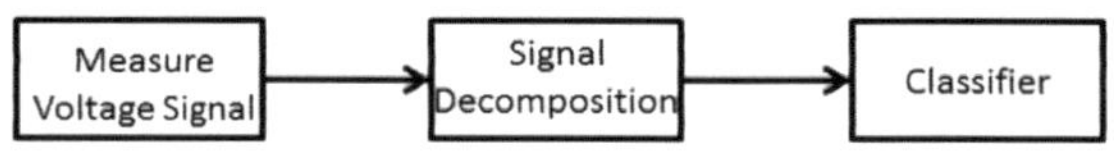

Figure 1: Block diagram of Fault detection in Microgrid

The detailed process of proposed methodology is discussed in flowchart diagram and it is shown in Fig. 2.

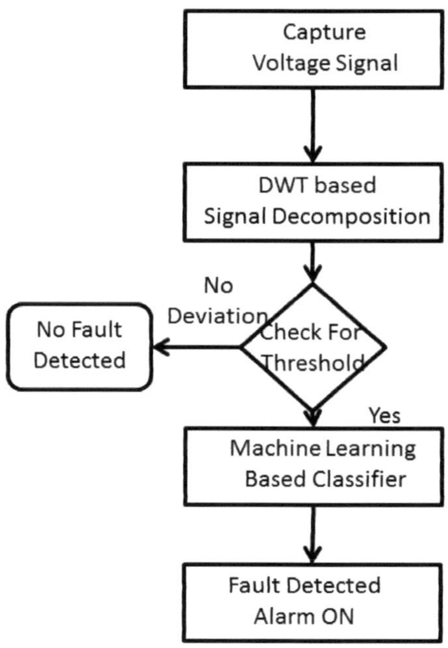

Figure 2: Flow chart of Proposed Methodology

In this proposed methodology voltage signal is decomposed up to 6 levels through discrete wavelet transform method. The details of decomposed signals are considered as data sets of no fault and arc fault signals. These data sets are trained in various Machine Learning classifiers for detecting the fault in AC microgrid.

2.1 Discrete Wavelet Transform (DWT)

It results in analyzing a signal into different frequencies at different resolutions, known as multi resolution. A wavelet is a wave-like oscillation that starts at zero, increases and decreases then come to starting point. It looks like as a "brief oscillation". Wavelet is our basic functions and acts as a window function. We can change the width of the wavelet and its central frequency as we move it across our signal by changing s. This is called scaling. We analyze wavelet coefficients at every possible scale produces a lot of data. This idea of choosing discrete values of dilation (a) and translation (b) parameters is implemented in,

i. Redundant Wavelet Transform (Frames) and

ii. Orthonormal bases for wavelets or Resolution Analysis (MRA)

2.2 Machine Learning Techniques

Machine Learning is set of techniques to make computers better at doing thing that humans can do better than machines. So, Machine Learning tries to make computers better at things where traditionally humans were performing better than machines. We make machines learn things like humans do and it is discussed in

2.2.1 Support Vector Machine (SVM)

It is the most effective classifiers among those, which has sort of linear. Support Vector Machines have a clever way to prevent over fitting. In Support Vector Machine, we want a classifier, so that it maximizes the separation of between the points and the decision surface. The distance of the closest point to the decision surface can be defined as the margin. The distance between the observations and the threshold is called a soft margin and it is discussed in [8].

Disadvantages of Support Vector Machine

1. Poor performance
2. SVMs do not provide probability estimates

2.2.2 Naïve Bayes

Naïve Bayes is a supervised machine learning method. It is also known as Gaussian and Kernel Naïve Bayes Method. It is mostly used for solving the classification problems effectively. It is called "naive" because it shapes the thing that is accepted as true the event of a certain feature is not depending on another for the event of other features. The Naïve Bayes is further discussed in [9].

2.2.3 Bagged Trees

Bagging is also known as bootstrapping aggregation. Bagging is most used procedure for reduction of the variance in statistical learning method. In other words, averaging a set of observations reduces the variance. In Bagging, we use row sampling with replacement method. Bootstrapping is a simple, powerful technique that is used a lot in both statistical inference and machine learning. The Bagged Trees further discussed in [10].

2.2.4 Decision Tree

A decision tree is a graphical representation. It is used instead of a table to show possibilities and results. Trees are more powerful than tables because table is limited to two dimensions and it is discussed in [11].

A decision tree consists of nodes and arcs:

Represents a decision nod□

Represents an outcome nodO

Indicates the possibilities or results \

It shows the order of decisions and outcomes.
Disadvantages of decision tree
1. Over fitting
2. High Variance
3. Low biased tree

2.2.5 K-nearest neighbor's (KNN)
It is one of the simplest Machine Learning algorithms. It stores all the available cases. In KNN, we categories cases based on a similarity measure then we calculate the distance of the nearest neighbor and it is more discussed in [12].

3. Results and Discussions

The Proposed Methodology is verified by IEEE9 bus benchmark micro grid test system. It consists of three power sources. One source is from Utility Grid of 18kv connected to bus number 2. The other source is PV based solar array of 16.5kv connected to Bus number 1. A diesel generator based source of 13.8kv connected at bus number 3. These sources are utilized by three loads, industrial load, health arena and residential load respectively. The Single line diagram of IEEE9 bus test system is shown in Fig.3.

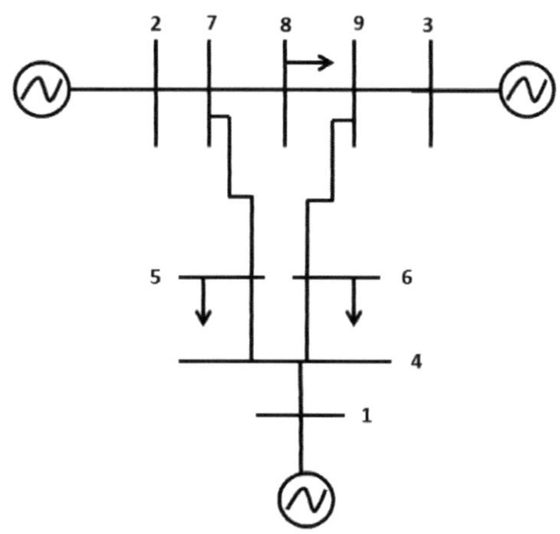

Figure 3: Single Line Diagram of IEEE9 Bus

The benchmark test system is developed and simulated in MATLAB/Simulink software. The test parameters like impedances are taken from the standard test system values. The line length is considered to be 2km. The load values MW, Mvar values are taken from the standard test system values. The MATLAB/Simulink diagram of IEEE9 benchmark microgrid test is shown in Fig.4.

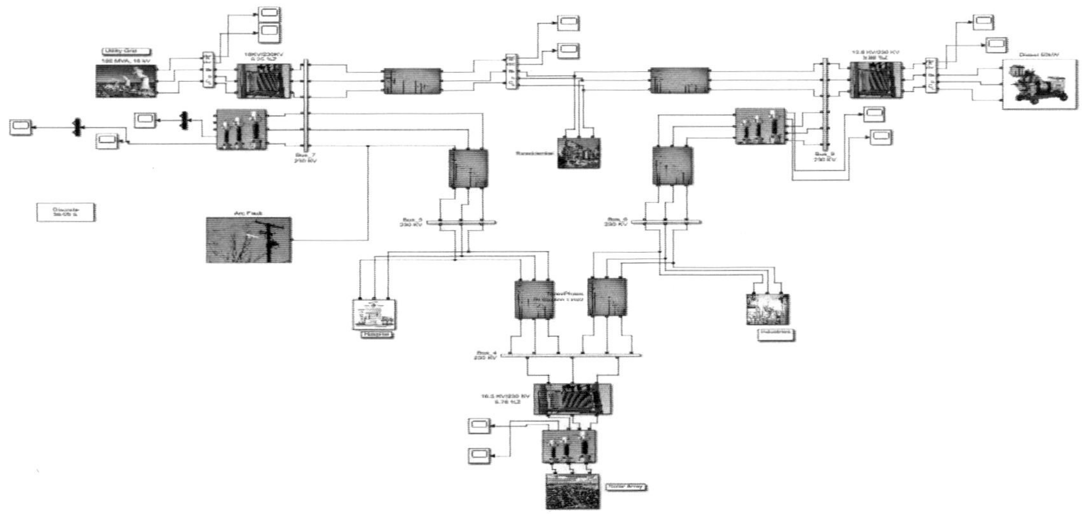

Figure 4: MATLAB/Simulink diagram of IEEE9 bus system

The test system is simulated and data collected at bus number 7. The Voltage Signal at bus number 7 without fault is shown in Fig.5.

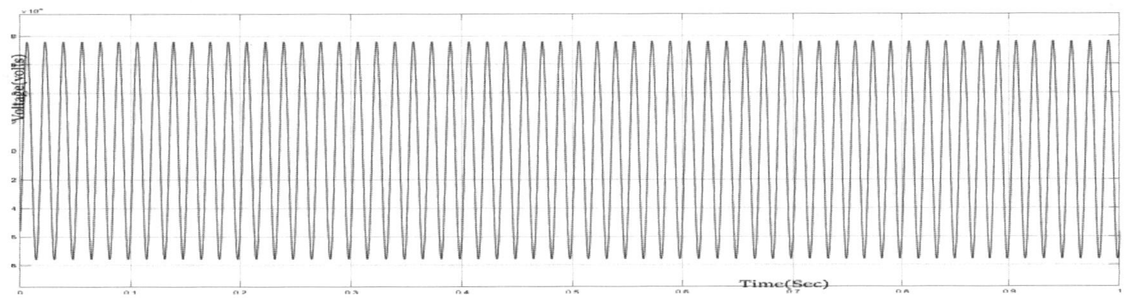

Figure 5: Voltage Signal at Bus Number 7

An Arc fault is modelled in MATLAB Simulink considering arc resistance and arc voltage. The Arc fault model is shown in Fig.6.

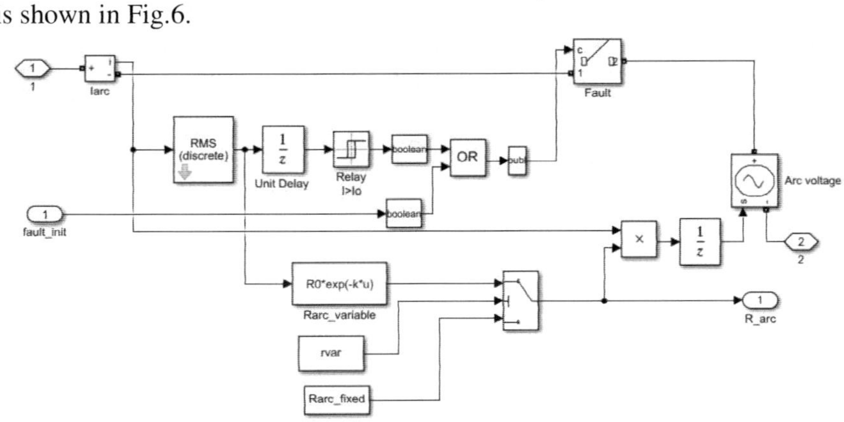

Figure 6: Simulink Diagram of Arc Fault

The arc fault is introduced to test system in between bus number 7 and bus number 5. The arc voltage at bus number 7 is now varied to actual original voltage as shown in the Fig.7.

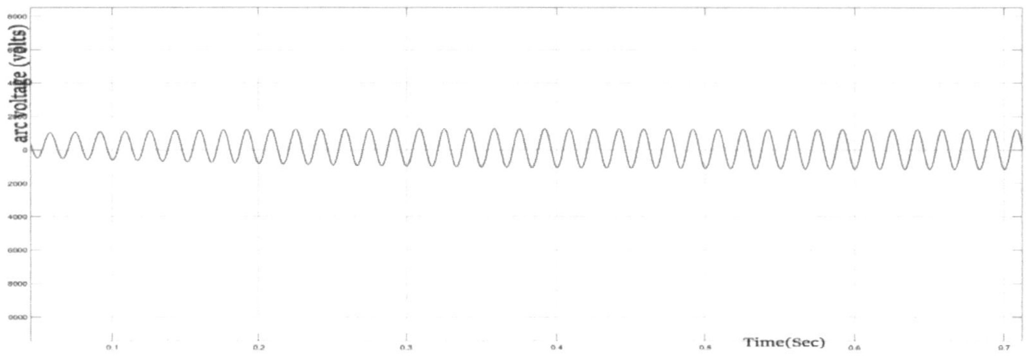

Figure 7: Arc Voltage at bus number 7

The no fault voltage and Arc Voltage signals are further processed in signal processing technique. In this paper discrete wavelet transform is considered as signal processing technique to decompose the no fault signal and arc fault signals. The window length of the signal decomposition is time period from 0.61 sec to 0.62 sec. The window length of the arc fault is shown in Fig.8. The total numbers of samples considered for decomposition are 201 samples each of arc fault and no arc fault.

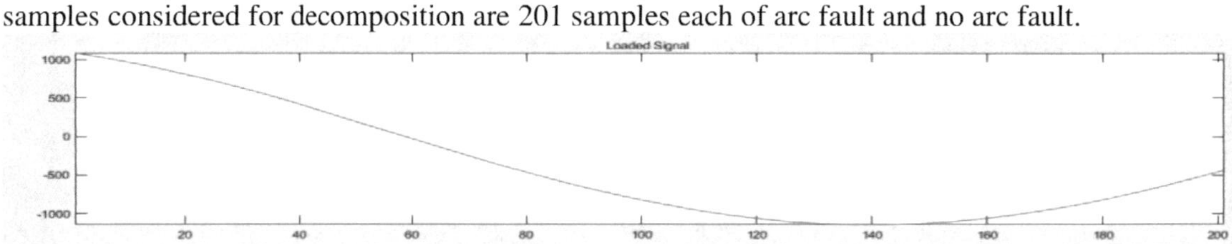

Figure 8: Window Length of arc fault signal

The Arc fault and no fault signals are decomposed up to 6 levels. The Daubechies (db.) wavelets are considered as mother wavelet. The detail coefficients are taken after decomposing the arc fault signal and no fault signal. The few detail signals at 6 levels for both no fault condition and arc faults are tabulated in Table 1.

Table 1: few sample details of arc and no fault condition

details1	details2	details3	details4	details5	details6	distortion
719.2638	-80.2801	-276.463	-535.313	-1019.29	-1367.27	nofaultA
-416.143	-312.366	-330.623	-550.831	-1023.62	-1361.13	nofaultA
2.156607	100.8254	237.8915	298.9359	202.3218	108.7423	nofaultA
-1.92877	341.1151	639.5612	916.8507	1098.612	1186.409	nofaultA
1.702905	-20.8173	874.3857	1302.913	1665.252	1871.87	nofaultA
5.822956	-0.79151	-2.80485	-6.57197	-16.9036	-42.4136	arcfaultA
-3.43142	-2.69734	-3.27784	-6.76196	-17.0603	-42.5736	arcfaultA
0.16589	1.115358	2.48717	3.157984	1.008335	-8.29741	arcfaultA
-0.16218	3.395779	6.580716	10.36898	14.1935	16.75161	arcfaultA
0.157528	-0.36881	9.002793	14.87102	22.49522	32.57347	arcfaultA

These details clearly show the variation between no fault detail values and arc fault details. It indicates that fault is detected. Now these data samples are trained in different machine learning techniques to detect the fault without any human supervision. It also can be studies the efficiency of various machine learning techniques in doing electrical fault detection problems. Since it is classification problem, we need confusion matrix to understand the behaviour of trained data and predicted data came after the raw data trained in the machine learning algorithm. Confusion matrix is a tool in data science to tell the performance of classification problem. It consists of True Values and False Values. The general confusion matrix for fault detection classification problem is shown in Table 2.

Table 2: General Confusion Matrix

Number of Samples	Fault	No Fault
Detection	True Fault (TF)	False Fault(FF)
No Detection	False No Fault(FNF)	True No Fault(TNF)

The accuracy of confusion matrix of classification model is calculated as $\frac{TF+TNF}{TF+FF+FNF+TNF}$ X 100. The Misclassification of Classifier model is calculated as $\frac{FF+FNF}{TF+FF+FNF+TNF}$ X 100. The details are trained in various machine learning techniques to study the performance of various machine learning techniques in solving fault detection classification Problem.

3.1 Decision Tree

The details data are trained in different types of decision tree machine algorithm, which is based on number of splits. In fine tree maximum numbers of splits are more than 100. In medium tree maximum numbers of splits are not more than 20. In Coarse Tree maximum numbers of splits are 4.The Confusion matrixes after training the data are shown in Fig.9.

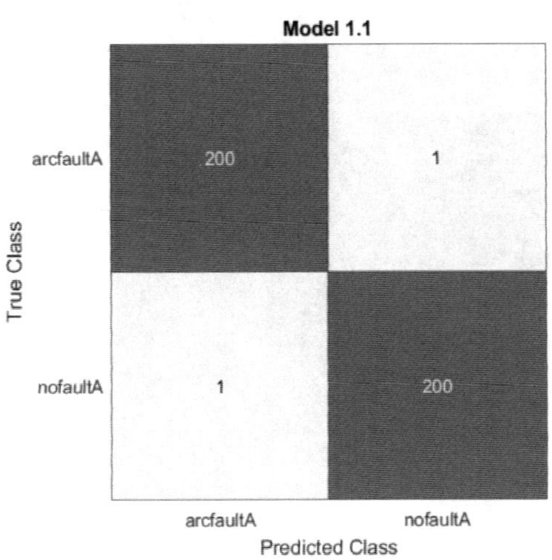

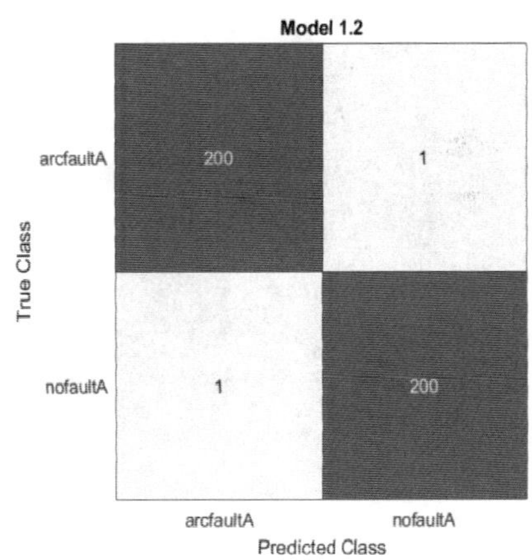

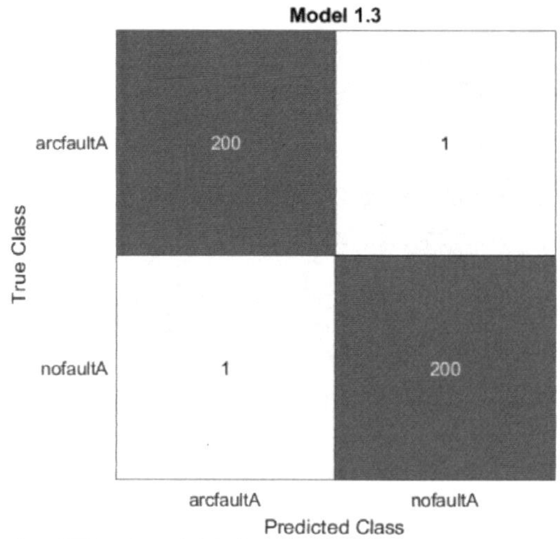

Figure 9: Confusion Matrix of Fine (model 1.1) Tree, Medium Tree (model 1.2) and Coarse Tree (model 1.3) Classification Model

In this machine learning algorithm, total 402 samples are trained. 201 samples are no fault samples and another 201 samples are arc fault samples. In Confusion matrix, out of 201 samples 200 samples are grouped as "true fault" and 1 sample is miss classified. The True fault detection percentage is 99.50% and misclassification percentage is 0.4975%. The total accuracy of the classification problem is 99.50%.

3.2 Naïve Bayes

The details data set is estimated in Naïve Bayes machine learning algorithm. 402 samples of arc fault and no fault data sets are predicted in Gaussian Naïve Bayes and Kernel Naïve Bayes machine learning techniques. The Confusion matrixes after training the data are shown in Fig.10.

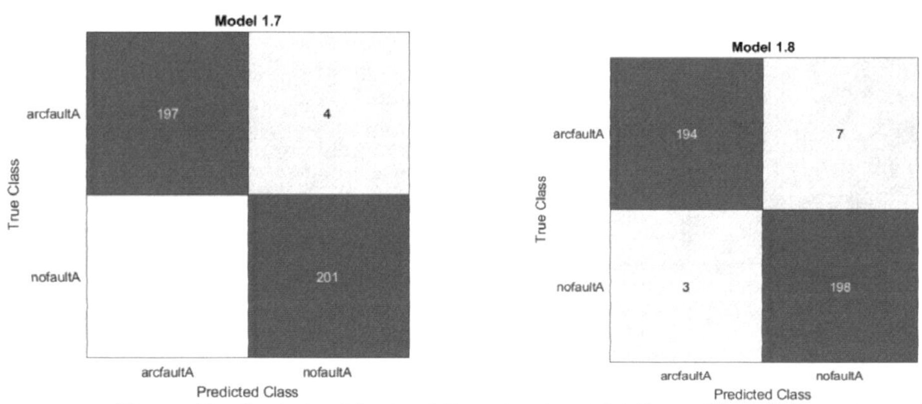

Figure 10: Confusion Matrix of Gaussian (model 1.7) and Kernel Naïve Bayes (model 1.8)

In Gaussian Naïve Bayes, out of 201 arc fault samples 197 samples are grouped as true fault and 4 samples are misclassified as no fault signals. The True fault detection percentage is 98.00% and misclassification percentage is 1.99%. The total accuracy of the classification problem is 99.00%. In Kernel Naïve Bayes, out of 201 arc fault samples 194 samples are grouped as true fault and 7 samples are misclassified as no fault signals. The True fault detection percentage is 96.50% and misclassification percentage is 3.48%. The total accuracy of the classification problem is 97.50%.

3.3 Support Vector Machine (SVM)

SVM machine learning classifier is classified according to the kernel function used in the objective function. In this Paper Cubic SVM, Various Gaussian, Optimizable SVM techniques performances are evaluated for fault detection classification problem. The Confusion matrixes after training the data are shown in Fig. 11.

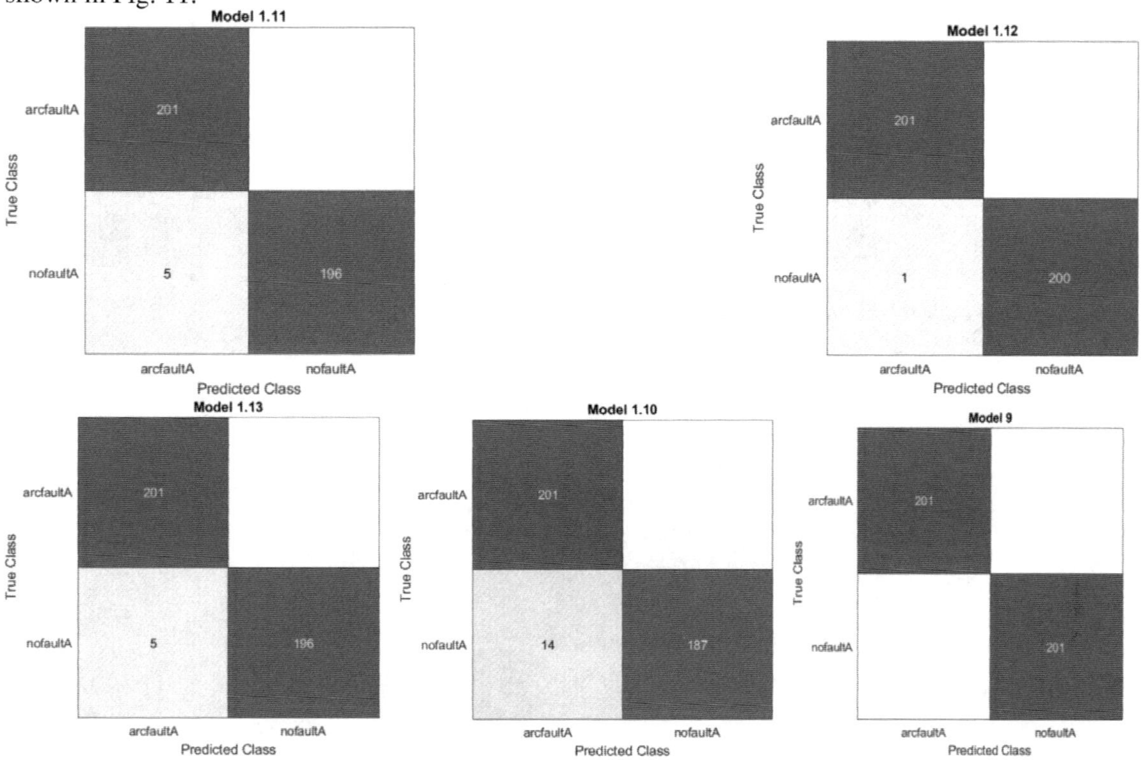

Figure 11: Cubic (model 1.11), Fine (model 1.12), Medium Gaussian (model 1.13), Quadratic (model 1.10), Optimizable SVM classifiers (model 1.9) Confusion Matrix

In Cubic SVM out of 201 arc fault samples 201 samples are grouped as true fault and 0 samples are misclassified as no fault signals. The True fault detection percentage is 100.00% and misclassification percentage is 0 %. The total accuracy of the classification problem is 98.75%. In Fine Gaussian SVM, out of 201 arc fault samples 201 samples are grouped as true fault and 0 samples are misclassified as no fault signals. The True fault detection percentage is 100% and misclassification percentage is 0%. The total accuracy of the classification problem is 99.75%.In Medium Gaussian SVM, out of 201 arc fault samples 201 samples are grouped as true fault and 0 samples are misclassified as no fault signals. The True fault detection percentage is 100 % and misclassification percentage is 0%. The total accuracy of the classification problem is 98.75%. In Quadratic SVM, out of 201 arc fault samples 201 samples are grouped as true fault and 0 samples are misclassified as no fault signals. The True fault detection percentage is 100% and misclassification percentage is 0%. The total accuracy of the classification problem is 96.50%. In Optimizable SVM, out of 201 arc fault samples 201 samples are grouped as true fault and no misclassified signals. The True fault detection percentage is 100% and misclassification percentage is 0%. The total accuracy of the classification problem is 100%.

3.4 k-nearest neighbors(KNN)

In this Paper Cubic KNN, Various Gaussian KNN, Optimizable KNN, Cosine KNN and weighted KNN techniques performances are evaluated for fault detection classification problem. In Cosine out of 201 arc fault samples 198 samples are grouped as true fault and 3 samples are misclassified as no fault signals. The True fault detection percentage is 98.75% and misclassification percentage is 1.49 %. The total accuracy of the classification problem is 93.78%. In Cubic KNN, out of 201 arc fault samples 201 samples are grouped as true fault and 0 samples are misclassified as no fault signals. The True fault detection percentage is 100% and misclassification percentage is 0%. The total accuracy of the classification problem is 97.00%.In Fine KNN, out of 201 arc fault samples 201 samples are grouped as true fault and 0 samples are misclassified as no fault signals. The True fault detection percentage is 100 % and misclassification percentage is 0%. The total accuracy of the classification problem is 99.75%. In Medium KNN, out of 201 arc fault samples 201 samples are grouped as true fault and 0 samples are misclassified as no fault signals. The True fault detection percentage is 100% and misclassification percentage is 0%. The total accuracy of the classification problem is 97.26%. In Weighted KNN, out of 201 arc fault samples 201 samples are grouped as true fault and no misclassified signals. The True fault detection percentage is 100% and misclassification percentage is 0%. The total accuracy of the classification problem is 97.76%. In Optimizable KNN, out of 201 arc fault samples 201 samples are grouped as true fault and 0 samples are misclassified as no fault signals. The True fault detection percentage is 100.00% and misclassification percentage is 0 %. The total accuracy of the classification problem is 99.75%.

3.5 Bagged Trees

The details data are trained in Bagged Trees machine algorithm, The Confusion matrixes after training the data are shown in Fig 12.

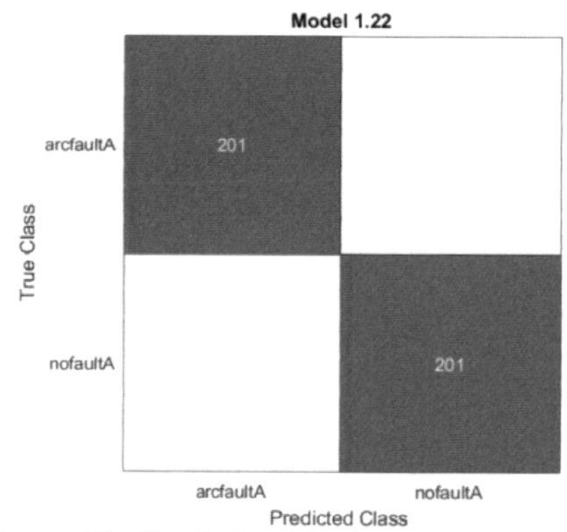

Figure 12: Confusion Matrix (Bagged Trees)

In Optimizable KNN, out of 201 arc fault samples 201 samples are grouped as true fault and 0 samples are misclassified as no fault signals. The True fault detection percentage is 100.00% and misclassification percentage is 0 %. The total accuracy of the classification problem is 100%.

4. Conclusions

From the Various machine learning classifiers, the observations are taken and tabulated in a Table 3.

Table 3: Overall Performances of Various Machine Learning Classifiers

Classifier	Fault Detection (Percentage)	Overall Classification Problem Accuracy (Percentage)
Fine Tree	99.50	99.50
Medium Tree	99.50	99.50
Coarse Tree	99.50	99.50
Gaussian Naïve	98.00	99.00

Bayes		
Kernel Naïve Bayes	96.50	97.50
Cubic SVM	100	98.75
Fine SVM	100	99.75
Medium SVM	100	98.75
Quadratic SVM	100	96.50
Optimizable SVM	100	100
Cosine KNN	98.75	93.78
Cubic KNN	100	97.00
Fine KNN	100	99.75
Medium KNN	100	97.26
Weighted KNN	100	97.76
Optimizable KNN	100	99.75
Bagged Trees	100	100

From the above Table 3 , It is concluded that Decision Tree learning classifier has more overall classification accuracy compared to SVM, KNN, Bagged Tree and Naïve Bayes learning classifiers. Whereas SVM , KNN, Bagged Trees have more fault detection accuracy than Naïve Bayes and Decision tree machine learning algorithm technique.

5. References

[1] S, Ramana Kumar Joga and Sinha, Pampa and Maharana, Manoj Kumar, Artificial Intelligence in Classifying High Impedance Faults in Electrical Power Distribution System (August 1, 2019). Proceedings of International Conference on Recent Trends in Computing, Communication & Networking Technologies (ICRTCCNT) 2019, Available at SSRN: https://ssrn.com/abstract=3430316 or http://dx.doi.org/10.2139/ssrn.3430316.

[2] Y. Bansal and R. Sodhi, "Microgrid fault detection methods: Reviews, issues and future trends," *2018 IEEE Innovative Smart Grid Technologies - Asia (ISGT Asia)*, Singapore, 2018, pp. 401-406, doi: 10.1109/ISGT-Asia.2018.8467938.

[3] J. J. Q. Yu, Y. Hou, A. Y. S. Lam and V. O. K. Li, "Intelligent Fault Detection Scheme for Microgrids With Wavelet-Based Deep Neural Networks," in IEEE Transactions on Smart Grid, vol. 10, no. 2, pp. 1694-1703, March 2019, doi: 10.1109/TSG.2017.2776310.

[4] 2 S. R. Fahim, S. K. Sarker, S. M. Muyeen, M. R. I. Sheikh, and S. K. Das, “Microgrid fault detection and classification: Machine learning based approach, comparison, and reviews,” *Energies*, vol. 13, no. 13, pp. 1–21, 2020.

[5] P. Kanakasabapathy and M. Mohan, "Digital protection scheme for microgrids using wavelet transform," 2015 IEEE International Conference on Electron Devices and Solid-State Circuits (EDSSC), Singapore, 2015, pp. 664-667, doi: 10.1109/EDSSC.2015.7285203.

[6] D. K. J. S. Jayamaha, N. W. A. Lidula and A. D. Rajapakse, "Wavelet Based Artificial Neural Networks for Detection and Classification of DC Microgrid Faults," 2019 IEEE Power & Energy Society General Meeting (PESGM), Atlanta, GA, USA, 2019, pp. 1-5, doi: 10.1109/PESGM40551.2019.8974108.

[7] M. F. Othman and H. A. Amari, "Online fault detection for power system using wavelet and PNN," 2008 IEEE 2nd International Power and Energy Conference, Johor Bahru, 2008, pp. 1644-1648, doi: 10.1109/PECON.2008.4762734.

[8] M. A. Hearst, S. T. Dumais, E. Osuna, J. Platt and B. Scholkopf, "Support vector machines," in IEEE Intelligent Systems and their Applications, vol. 13, no. 4, pp. 18-28, July-Aug. 1998, doi: 10.1109/5254.708428.

[9] F. Yang, "An Implementation of Naive Bayes Classifier," 2018 International Conference on Computational Science and Computational Intelligence (CSCI), Las Vegas, NV, USA, 2018, pp. 301-306, doi: 10.1109/CSCI46756.2018.00065.

[10] S. Clémençon, M. Depecker and N. Vayatis, "Bagging Ranking Trees," 2009 International Conference on Machine Learning and Applications, Miami Beach, FL, 2009, pp. 658-663, doi: 10.1109/ICMLA.2009.14.

[11] A. Navada, A. N. Ansari, S. Patil and B. A. Sonkamble, "Overview of use of decision tree algorithms in machine learning," 2011 IEEE Control and System Graduate Research Colloquium, Shah Alam, 2011, pp. 37-42, doi: 10.1109/ICSGRC.2011.5991826.

[12] S. Zhang, X. Li, M. Zong, X. Zhu and R. Wang, "Efficient kNN Classification With Different Numbers of Nearest Neighbors," in IEEE Transactions on Neural Networks and Learning Systems, vol. 29, no. 5, pp. 1774-1785, May 2018, doi: 10.1109/TNNLS.2017.2673241.

IOP Conf. Series: Materials Science and Engineering 1131 (2021) 012013 doi:10.1088/1757-899X/1131/1/012013

Performance evaluation of health recommendation system based on deep neural network

Gauri Sood[1], Neeraj Raheja[2]

[1]Research Scholar, CSE Department, Maharishi Markandeshwar (Deemed to be University), Mullana 133207, India

[2]Associate professor, CSE Department, Maharishi Markandeshwar (Deemed to be University), Mullana 133207, India

er.gauri31sood@gmail.com

Abstract. Deep learning has developed as an innovative zone of machine learning and data mining exploration part. Controlled or unconfirmed methodologies which contain of a number of layers of handling which form a hierarchy are castoff for preparation in deep learning. Every succeeding layer mines an ever more intellectual depiction of the input data and shapes upon the depiction from the preceding layer, usually by calculating a nonlinear alteration of its input. The constraints of these alterations are adjusted by preparation of the prototypical on a dataset. A deep learning prototypical studies better depiction as it is delivered with more volumes of data. Key objective of using deep learning methods in recommender schemes is to lower time complexity and to increase the accurateness of formed expectations. In this paper, performance of planned HRS is evaluated by Arbitration Time, Latency Time, Jitter, Execution Time, Network Bandwidth Consumption, Power Consumption, Training Accuracy and Testing Accuracy.

1. Introduction

The explosive progress in the work of digital data and large count of people on internet has produced an impending task of data burden that obstructs well-timed entrance to things of importance on the internet. Recommender schemes are facts purifying arrangements which contract with the difficulty of data excess by clarifying vibrant evidence portion out of huge volume of vigorously produced data allowing to customer's favorites, awareness, or practical activities around point. Next generation databanks and data learning implements are necessary due to rapid progression of statistics and information. Every business wants a service recommendation arrangement which can be used by many of customers. The number of consumers, goods and data is growing very speedily; service recommender systems face the big data exploration problematic. Traditional recommender facility arrangements suffer from shortage of scalability and effectiveness difficulties while handling or examination of this statistics on outsized scales [1]. A recommender arrangement must fulfill a right level of estimate correctness to develop desire capability and efficiency.

Content from this work may be used under the terms of the Creative Commons Attribution 3.0 licence. Any further distribution of this work must maintain attribution to the author(s) and the title of the work, journal citation and DOI.
Published under licence by IOP Publishing Ltd

1.1Overview of deep learning

Deep learning is referred as subclass of simulated neural systems has many management stages. Anyhow algorithmic progressions, the development in handing out capabilities, exploiting GPUs and the assortment of better-quality datasets are generally factors which assisted in the ongoing flood of deep knowledge.

1.1.1What is deep learning?

- It comes under class of machine learning systems
- It uses hierarchy of nonlinear processing layers and complex model structures
- Layers learn to represent different representation of facts
- Advanced stage features are constructed from minor stage abstract features
- Trendy name for "Neural Networks with deep layers"

Deep learning has become increasingly more famous all through subfields of software engineering, for example, Natural language preparing, picture and video handling, PC visualization, and information withdrawal because there has not been such a typical way to deal with in taking care of various types of figuring issues previously. With such part of profound learning systems, they are not just exceptionally equipped for helping complex issues in numerous fields; however they additionally structure a mutual terminology and shared view for these exploration grounds. Deep learning strategies also assist these sub-fields to team up with one another wherever this was somewhat questionable earlier because of the assorted variety and unpredictability of used systems [2].

For whatever length of time that the personalization pattern stays famous, the recommender frameworks research will assume a basic job in data examining. In spite of the fact that the use of profound learning into recommender frameworks field guarantees noteworthy and empowering results, difficulties, for example, the precision and marketability are as yet open for upgrades and warrant future work.

1.1.2 Why Deep Learning for Search and Recommender System?

- Direct content Feature extraction instead of metadata
- Text, Image, Audio
- Better representation of users and items
- Hybrid algorithms and heterogeneous data can be used
- Better suited to model dynamic behavioral patterns and complex feature interactions.

1.1.3 Health Recommendation System

A health recommender system (HRS) is one of a specialization of a recommender system. The aim of an HRS is to deliver medical information to it's user which is meant to be extremely pertinent to the medical enhancement of the associated patient.

2. Related work

Baldominos et al. [3] presented Data Care, a resolution for intellectual wellbeing administration. This creation is intelligent to recover and collect facts from dissimilar fundamental presentation signs in wellbeing hubs, however too to guess upcoming ideals for these strategic presentation signs and, by way of consequence, fervor initial warnings once objectionable standards are nearly to happen or offer endorsements to increase the excellence of facility.

Belle et al. [4] discussed few of key challenges with an emphasis on three future and encouraging regions of medical examination: image, sign, and genomics based analytics and current research which marks use of huge capacities of medicinal data while joining multimodal facts from dissimilar sources.

Georgia Koutrika [5] enclosed current progresses in recommendation techniques, concentrating on matrix factorization, multi-armed thieves, and approaches for merger recommendations. The author also described evaluation techniques, and outline open problems and challenges.

Konstan and Riedl [6] presented an analysis on important advances in collaborative filtering recommender schemes, concentrating on the development from investigation focused only on processes to investigation focused on the ironic set of queries around the customer involvement with the recommender. Author assessed the customer understanding of a recommender needs a wider set of actions that have remained normally castoff, and recommended extra actions that must confirmed operative.

Zhang et al. [7] projected a prototypical merging a collective clarifying commendation procedure with profound knowledge skill. These prototypical practices a feature demonstration scheme built on a quadric polynomial reversion prototypical, which gets hidden topographies further precisely by successful upon the old-style matrix factorization procedure. Formerly, these hidden features are considered as the contribution records of the profound neural system prototypical, and are castoff to forecast the assessment totals.

Riyazand Varghese [8] applied an innovative recommendations scheme by collaborative filtering procedure in Apache Hadoop leveraging Map Reduce model for giant data. Apache Hadoop is an exposed structure for scattered dealing out arrangements and can route huge capacities of records. This can be castoff for disconnected handling and inefficient for small inactivity analytics. This harbor statistics on top of the subsequent collection catalogues alike HBase and enhance the presentation of it. For the creation commendations the Amazon dataset is castoff. Planned Structure has substantial upgrading 3 in performance compared to conventional implements.

Jooa et al. [9] planned and applied a recommendation scheme which examines by configurations and individual tendencies of consumers by exhausting relationship instruction examination and collaborative filtering for composed client statistics on staying customer businesses with NFC. The commendation procedure castoff in planned scheme castoff the statistics examination outcomes plus remoteness facts from GPS to mention native productions that persons are greatly expected to stay.

Sahu et al. [10] exploited Content Centered Clarifying, Collaborative Grounded Clarifying, Mixture Contented-Cooperative, k-mean grouping and Naive Bayes categorizer to their great in direction to attain the finest conceivable exactness and ensure a complete reasonable examination. The strong point of altogether procedures can be obviously understood by the noteworthy improvement in accurateness, showed by the investigational exploration compelling unkind start problematic into attention.

Luis et al.[11]examined the probability of construction a content centered reference scheme which associates wellbeing customers to trustworthy fitness informative web links from Medline Plus for a particular well-being audio visual from YouTube. The database for this training contains a gathering of fitness linked audiovisual and their obtainable metadata. Semantic tools were castoff to endorse wellbeing web links from Medline Plus.

Abdiet al. [12] showed an intensive literature analysis to classify research documents on Recommender Schemes in the Healthcare in the last ten years. The important conclusions of this review are that the submission of Recommender Systems in Healthcare is growing. The integration of appropriate data is restricted although it has been recommended as a key element to refining the superiority of the recommendations and the accurateness of the predications.

Wang et al. [13] presented the Content centered Papers& Consultations Recommender Arrangement on computer discipline and its network facility at http://www.keaml.cn/prs/. This scheme mentions appropriate papers or consultations through a precedence direction grounded on the intellectual of a manuscript.

Ayyaz et al. [14] suggested a Mixture Content centered Indistinct Conformal Recommender Arrangement. Content centered clarifying system with fuzzy interpretation system. The author used two datasets Movie Lens and Movie Tweetings for experiment to recommend shows to the customers and assessed and then matched with further advanced reference schemes. The suggested procedure provided improved outcome than the old-style approaches.

Fazazi et al. [15] presented a recommender e-learning methodology which make use of commendation practices for informative data taking out precisely for recognizing e-Learners' education inclinations. The suggested method is centered on three components, a field component which covers completely the information for a specific zone, a beginner component that practices to detect beginners' education likings and actions and a commendation unit which practices records to make an appropriate reference list and guessing performances.

Batmaz et al. [16] analyzed collected readings in four extents which are profound knowledge prototypes used in reference arrangements, preparations for the tests of reference arrangements, attentiveness and dominance above endorsement fields, and the purposive assets. Author too provided a complete quantifiable valuation of journals in the field and concludes by deliberating added visions and conceivable upcoming effort on the topic.

Sahoo et al. [17] proposed smart HRS by means of (RBM)-(CNN) profound knowledge technique, which delivers an vision into in what way large statistics analytics can be castoff for the execution of an operative fitness reference device, and explains an prospect for the wellbeing maintenance productiveness to changeover from an old-style situation to a further adapted standard in a tele-health situation.

Nouh et al. [18] suggested a nifty recommender arrangement built on the procedures of cross knowledge for individual comfort facilities, named a clever reference scheme of mixture knowledge. The important fitness feature is deliberated to be individual routine, with the benefit of a life threatening inspection of several restraints. Incorporating the reference scheme efficiently gives to the avoidance of sickness, and it too clues to a decrease in cure fee, which adds to an enhancement in excellence of lifecycle.

3. Proposed work

Deep learning has developed as an innovative zone of machine learning and data mining exploration part. Controlled or unconfirmed methodologies which contain of a number of layers of handling which form a hierarchy are castoff for preparation in deep learning. Every succeeding layer mines an ever more intellectual depiction of the input data and shapes upon the depiction from the preceding layer, usually by calculating a nonlinear alteration of its input. The constraints of these alterations are adjusted by preparation of the prototypical on a dataset. A deep learning prototypical studies better depictions as it is delivered with more volumes of data. Key objective of using deep learning methods in recommender schemes is to lower time complexity and to increase the accurateness of formed expectations. Figure 1 demonstrates the process of machine learning.

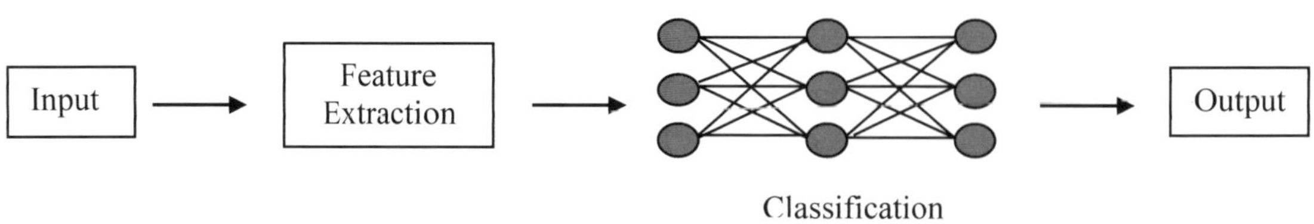

Figure 1. Machine Learning

Extracting features manually from a dataset, strong knowledge of the subject as well as the domain is necessary. It is an extremely time-consuming process. With Deep Learning, we can automate the process of Feature Engineering as shown in figure 2.

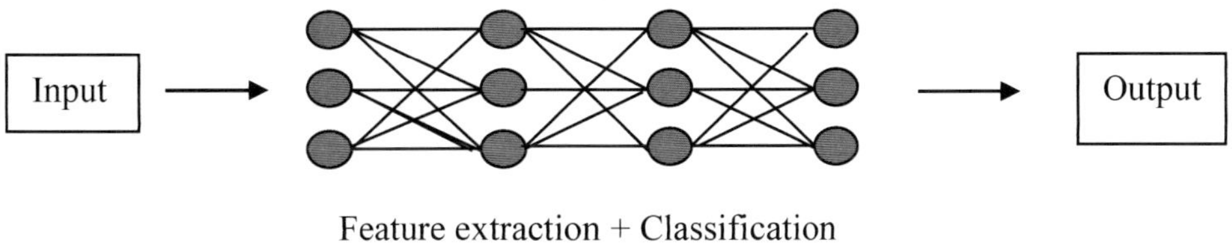

Figure 2. Deep Learning

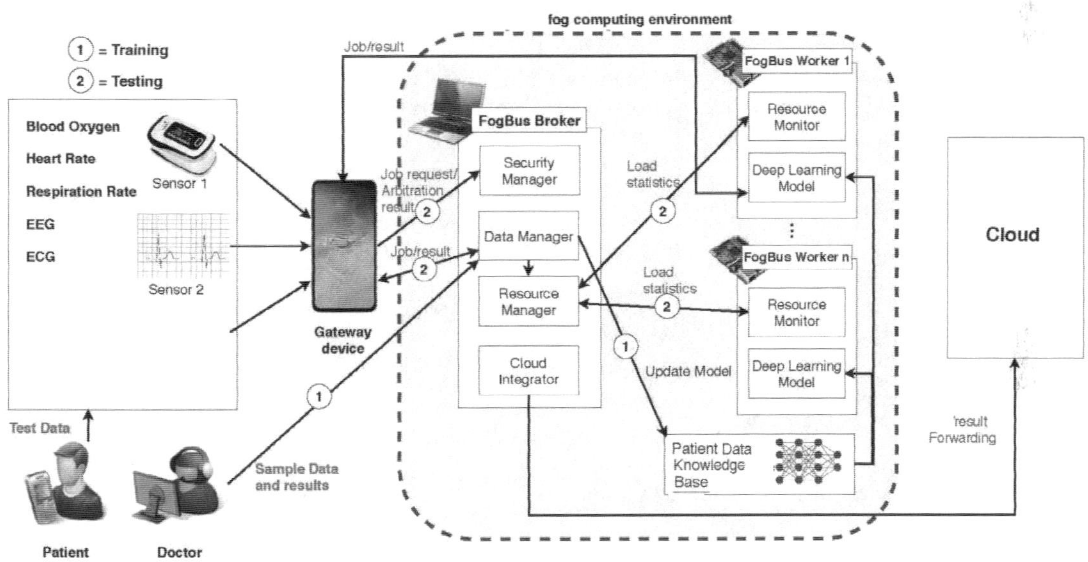

Figure 3. Proposed Architecture

In this paper, deep neural network architecture for recommender scheme is designed as shown in figure 3. In feature extraction, all the required features are extracted and in feature selection, the important features that improve the performance of deep learning model are selected. The neural network is used for classification. The multilayer perceptron (MLP) is used to categorize linearly attached data. The MLP neural networks are feed forward networks with one or more layers of components between the input and output layer. The output components signify a hyper plane in the input space. In this the training procedure of the MLP neural networks uses the back propagation (BP) algorithm based on the generalized delta rule. The BP algorithm is enhanced by relating the momentum technique which suggests addition a span α with a worth between (0, 1] to alter the weights. A typical value for constraint α is 0.8. The back Propagation procedure with

IOP Conf. Series: Materials Science and Engineering 1131 (2021) 012013 doi:10.1088/1757-899X/1131/1/012013

gradient and momentum has a comparable intricacy to the BP algorithm but is improved concerning the merging speed and the capability to evade local minima.

3.1 Dataset used

In this paper, the proposed recommendation methodology is evaluated on real datasets [19].

3.1.1 Tool used

In this paper all evaluations are performed using Python 3.10. Highly-productive interface Keras are used for solving machine learning complications. Keras is a deep learning API written in Python, running on top of the machine learning platform. It is extensively used to speedily develop new training techniques or striking model architectures. In this paper, proposed churn prediction technique in Python programming language and Keras library for planned deep neural network architectures is implemented.

Table 1: Evaluation Parameters

Name	Description
Arbitration Time	It is the time to settle the differences between client and Cloud.
Latency Time	Time interval between a client request and cloud service provider's response.
Jitter	Latency variability, it is a measure of network and CPU congestion.
Execution Time	It is completion time of the task.
Network Bandwidth Consumption	It is the capacity of a wireless network communications to transmit the maximum amount of data from one point to another over an internet connection in a given amount of time.
Power Consumption	It is energy intake by device per unit time.
Training Accuracy	It is the accuracy when applying the model on training data.
Testing Accuracy	It is the accuracy when applying the model for testing data.

4. Results and discussion

In this paper, performance is evaluated of planned HRS by Arbitration Time, Latency Time, Jitter, Execution Time, Network Bandwidth Consumption, and Power Consumption. Table 2 and figure 4 demonstrates arbitration time in different scenarios.

Table 2. Arbitration Time in different scenarios

	Master Node	Node-1	Node-2	2 -Nodes-Ensemble	Cloud
1	26	1377	3034	2122	61
2	39	1322	2799	2806	53
3	211	1456	3132	3115	89
4	179	1493	2954	4150	121
5	165	1411	2910	3315	254
Avg	124	1411.8	2965.8	3101.6	115.6

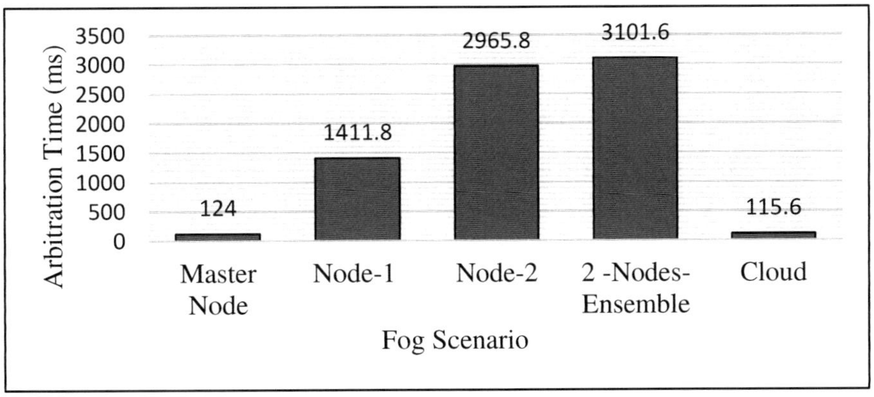

Figure 4. Arbitration Time in different scenarios

- Table 3 and figure 5 demonstrate latency in different scenarios.

Table 3. Latency in different scenarios

	Master Node	Node-1	Node-2	2 -Nodes-Ensemble	Cloud
1	10	27	19	27	2585
2	11	25	17	34	2750
3	14	21	24	23	2650
4	25	19	19	25	2555
5	21	17	20	36	2672
Avg	16.2	21.8	19.8	29	2642.4

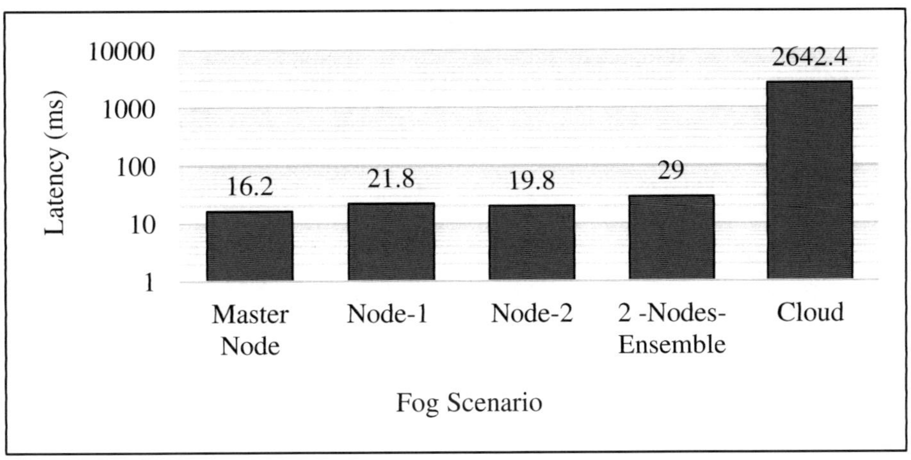

Figure 5. Latency in different scenarios

- Table 4 and figure 6 demonstrates jitter in different scenarios.

Table 4. Jitter in different scenarios

	Master Node	Node-1	Node-2	2 -Nodes-	Cloud

				Ensemble	
1	1	2	2	7	165
2	3	4	7	11	100
3	11	2	5	2	95
4	4	2	1	11	117
Avg	4.75	2.5	3.75	7.75	119.25

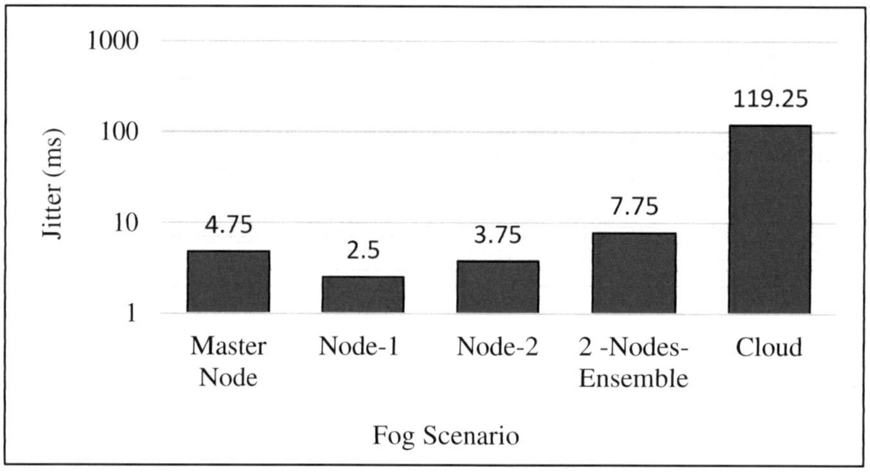

Figure 6. Jitter in different scenarios

- Table 5 and figure 7 demonstrate total execution time in different scenarios.

Table 5. Total Execution Time in different scenarios

	Master Node	Node-1	Node-2	2-Nodes-Ensemble	Cloud
1	2215	3718.5	4944	6307.5	242
2	2315	3709.5	3637.5	6277.5	391
3	2490	3757.5	3408	6082.5	512
4	2450	3744	3307.5	5955	1600
5	2420	3784.5	3381	4492.5	1028
Avg	2378	3742.8	3735.6	5823	754.6

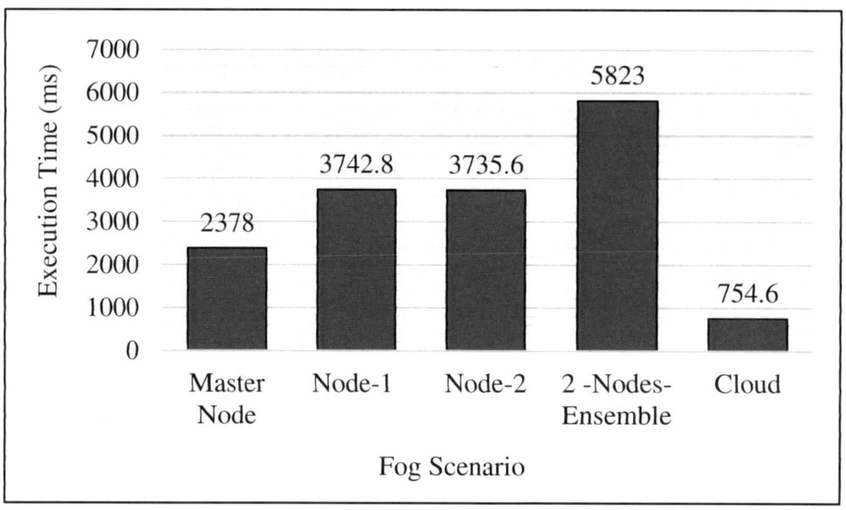

Figure 7. Execution Time in different scenarios

- Table 6 and figure 8 demonstrate network bandwidth in different scenarios.

Table 6. Network Bandwidth in different scenarios

	Master Only	Node-1	Node-2	2-Nodes-Ensemble	Cloud
1	4.1	5	13	19	18
2	4.1	5	14	16	17
3	4.1	5	25	18	21
4	4.1	6	8	29	19
5	4	7	9	21	14
Avg	4.08	5.6	13.8	20.6	17.8

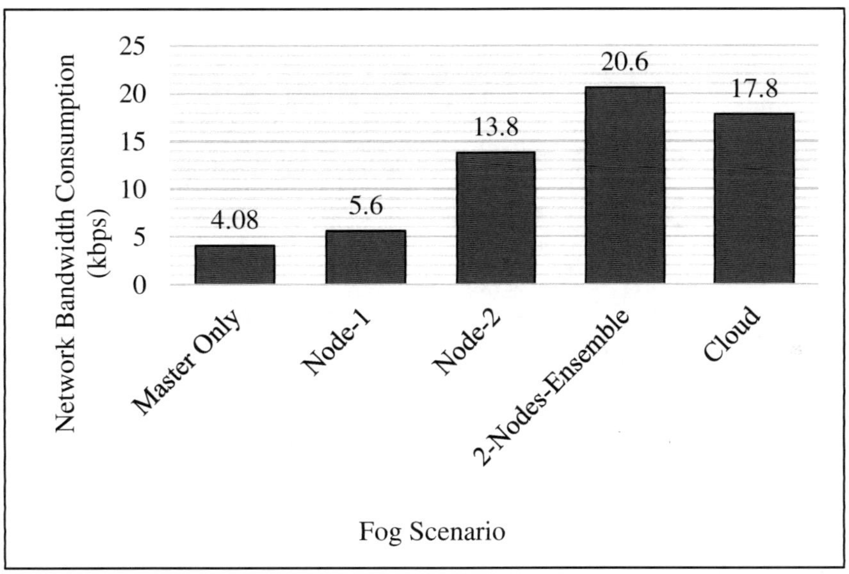

Figure 8. Network Bandwidth in different scenarios

- Table 7 and figure 9 demonstrate energy (power) consumption in different scenarios.

Table 7. Energy in different scenarios

Master Node	Node-1	Node-2	2-Nodes-Ensemble	Cloud
2.75	2.83	3.4422	4.1545	17.935

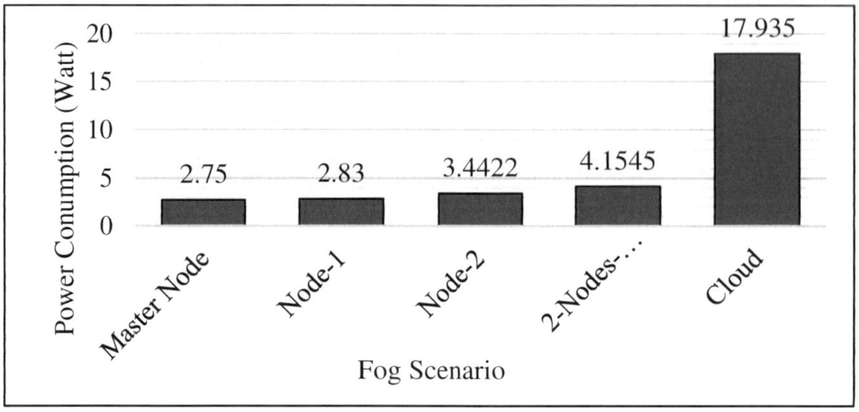

Figure 9. Energy in different scenarios

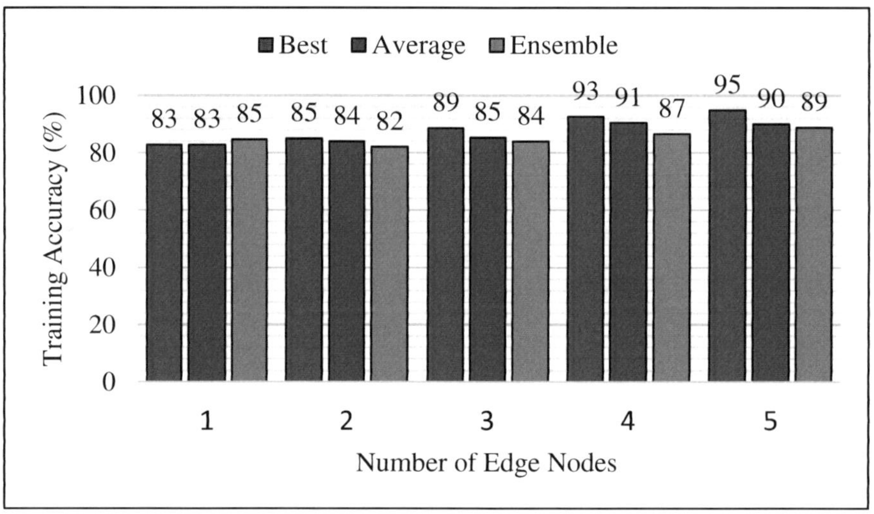

Figure 10. Training Accuracy with number of edge nodes

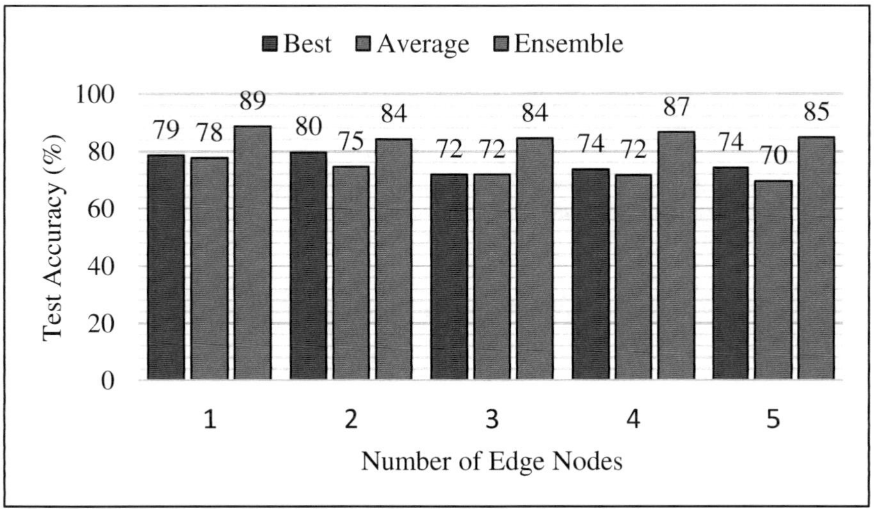

Figure 11. Network Bandwidth with number of edge nodes

- Figure 10 shows Training Accuracy and Figure 11 shows Testing Accuracy of HRS.

5. Conclusion

Simple assessment rules centered and complex grouping models for healthcare recommendation tasks have been discussed in the literature. These approaches are capable to execute recommendation jobs but make use of manual feature selection and are more time consuming and error-prone. In this paper, deep neural network based healthcare recommendation system is created to automate feature selection engineering process. Experiment is conducted using real world datasets.

6. Future scope

Other learning techniques can also be used as upcoming research to improve the performance. In addition, hybrid procedures can be used to offer more assessment.

References

[1] A Adomavicius, G., & Tuzhilin, A. (2005). Toward the next generation of recommender systems: A survey of the state-of-the art and possible extensions. IEEE transactions on knowledge and data engineering, 17(6), 734-749.

[2] Chen, C., Qin, C., Qiu, H., Tarroni, G., Duan, J., Bai, W., & Rueckert, D. (2020). Deep learning for cardiac image segmentation: A review. *Frontiers in Cardiovascular Medicine*, *7*, 25.

[3] Baldominos, A., De Rada, F., & Saez, Y. (2018). DataCare: Big Data Analytics Solution for Intelligent Healthcare Management. International Journal of Interactive Multimedia & Artificial Intelligence, 4(7).

[4] Raghupathi, W., & Raghupathi, V. (2014). Big data analytics in healthcare: promise and potential. Health information science and systems, 2(1), 3.

[5] Koutrika, G. (2018). Recent Advances in Recommender Systems: Matrices, Bandits, and Blenders. In EDBT (pp. 517-519).

[6] Mahrishi, M., Hiran, K. K., Meena, G., & Sharma, P. (2020). Machine Learning and Deep Learning in Real-Time Applications. IGI Global.

[7] Zhang, L., Luo, T., Zhang, F., & Wu, Y. (2018). A recommendation model based on deep

neural network. IEEE Access, 6, 9454-9463.
[8] Riyaz, P. A., & Varghese, S. M. (2016). A scalable product recommendations using collaborative filtering in hadoop for bigdata. Procedia Technology, 24, 1393-1399.
[9] Jooa, J., Bangb, S., & Parka, G. (2016). Implementation of a recommendation system using association rules and collaborative filtering. *Procedia Computer Science*, *91*, 944-952.
[10] Sahu, S. P., Nautiyal, A., & Prasad, M. (2017). Machine Learing Algorithms for Recommender System–a comparative analysis. International Journal of Computer Applications Technology and Research, 6(2), 97-100.
[11] Bocanegra, C. L. S., Ramos, J. L. S., Rizo, C., Civit, A., & Fernandez-Luque, L. (2017). HealthRecSys: A semantic contentbased recommender system to complement health videos. BMC medical informatics and decision making, 17(1), 1-10.
[12] Mohamed H. A., George O. O., Ronald W.M. (2017). A Survey of Context-Aware Healthcare Recommender Systems. International Journal of Science and Research (IJSR), 962-967.
[13] Wang, D., Liang, Y., Xu, D., Feng, X., & Guan, R. (2018). A content-based recommender system for computer science publications. Knowledge-Based Systems, 157, 1-9.
[14] Ayyaz, S., Qamar, U., & Nawaz, R. (2018). HCF-CRS: A Hybrid Content based Fuzzy Conformal Recommender System for providing recommendations with confidence. PloS one, 13(10), e0204849.
[15] El Fazazi, H., Qbadou, M., Salhi, I., & Mansouri, K. (2018). Personalized recommender system for e-Learning environment based on student's preferences. INTERNATIONAL JOURNAL OF COMPUTER SCIENCE AND NETWORK SECURITY, 18(10), 173-178.
[16] Batmaz, Z., Yurekli, A., Bilge, A., & Kaleli, C. (2019). A review on deep learning for recommender systems: challenges and remedies. Artificial Intelligence Review, 52(1), 1-37.
[17] Sahoo, A. K., Pradhan, C., Barik, R. K., & Dubey, H. (2019). DeepReco: deep learning based health recommender system using collaborative filtering. Computation, 7(2), 25.
[18] Nouh, R. M., Lee, H. H., Lee, W. J., & Lee, J. D. (2019). A smart recommender based on hybrid learning methods for personal well-being services. Sensors, 19(2), 431.
[19] https://github.com/Cloudslab/HealthFog/blob/master/HeartMo del/data.xlsx.

Comparative Study of Recent CMFD Techniques: A Review

N Parashar[1]

[1]Department of Computer Engineering and Applications, GLA University Mathura, Uttar Pradesh, India

E-mail: nishtha.parashar@gla.ac.in

Abstract. Due to the ease in manipulation of digital images with various existing photo editing tools and softwares the task of authentication and identifying the reliability of digital image has become a major concern. Copy-move forgery detection (CMFD) is therefore known to be one of the key-domains in the latest digital image authentication study. Copy-move forgery is a passive approach in which a portion of an image is copied and then pasted onto the same image, resulting in a tampered image. In this paper, current developments in CMFD have been surveyed and a comparative evaluation of recent CMFD techniques has been done along with its advantages and limitations. Also, a detailed description of relevant copy-move forgery detection datasets is provided which will help the researchers to decide which dataset to choose for a given CMFD approach.

1. Introduction

In today's digital world, images are considered as one of the significant means of communication among the people for interaction. Image authentication thus serves as a means of preventing digital images from being misused for unwanted use. It helps in ensuring that the given image is original without any manipulations. Thus, active approach and passive approach are the two approaches used in digital image forensics [1]. Digital watermarking is considered as one of the well-known active approach in which embedding of watermark is performed on the host image. However, the drawback here is that it needs to be embedded in advance on the given image. Therefore, passive methods such as copy-move forgery detection help to address this necessity of pre-processing. Figure 1 demonstrates the detailed classification of both active and passive areas of image forgery detection.

This paper aims to present, together with their advantages and limitations, a comparative evaluation of recent copy-move forgery detection techniques. Also, the detailed list of publicly available datasets is provided which will help the researchers in deciding suitable dataset for the given copy-move forgery detection approach.

Content from this work may be used under the terms of the Creative Commons Attribution 3.0 licence. Any further distribution of this work must maintain attribution to the author(s) and the title of the work, journal citation and DOI.
Published under licence by IOP Publishing Ltd

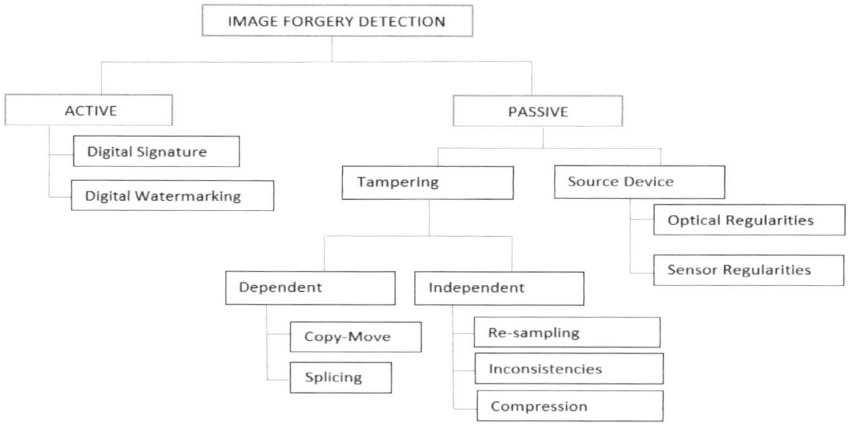

Figure 1. Classification of image forgery detection

2. Copy-Move Forgery

Copy-move forgery detection is a passive technique that involves copying and pasting different image areas onto the same image. Figure 2 contrasts an original image with its tampered counterpart produced by copy-move forgery [2]. At an initial stage, pre-processing of image is being done after that feature extraction is performed and then similarity match between features is accomplished. Figure 3 illustrates a generalised copy-move forgery detection (CMFD) workflow.

The numerous techniques used to detect copy-move forgery can be divided into two categories: block-based methods and key-point-based methods. In block-based methods, input image regions are analyzed to find candidate matches, while correspondence between points is found in key point-based methods [1] [7] [17].

Figure 2. (a) Original Image (b) Forged Image

Figure 3. Generalized CMFD Process

3. Related Work

Due to lack of prior knowledge about the authenticity of the original image forensic methods are considered as blind in nature. Currently, various algorithms have been proposed by the researchers in the domain of detecting copy-move forgery.

J. Zhong et al. [1] developed a two-pass hash function representation and search method to improve the good score and high efficiency in detecting copy-move forgery. X. Wang et al. [2] introduced an adaptive key-point based method to find robust invariant features and screen out the wrong pairs. This method helps detect the use of copy-move forgery in a fast and effective manner. Another effective approach premised on singular decomposition of value and Kolmogorov Smirnov test was proposed by B. Ahmed et al. [8] which helped in the reduction of false positive rate. G. Gani et al. [10] put forward a DCT and cellular automata-based approach for detecting forgeries under post processing attacks. In this, feature vectors were constructed using cellular automata based on the DCT coefficients. K.B. Meena et al. [11] suggested a tetrolet transform-based CMFD technique that increased accuracy considerably. A tetrolet transform based CMFD technique was proposed by K.B. Meena et al. [11] which improved the accuracy. Y. Zu et al. [12] introduced AR-Net based end to end neural network for enriching the feature representation. C. Chen et al. [3] proposed SIFT based invariant moments calculation and region growing strategy for identifying regions prone to copy-move forgery. Hybrid local feature extraction-based technique was developed by B. Soni et al. [4] which was an enhancement in CMFD methods. H. Huang et al. [5] introduced a key-point based method for finding relevant forensic information from an image using super pixel resolution and Helmert transformation. B. Elhaminia et al. [6] used the concept of Markov Random Field to deal with copy-move forgery detection and achieved an improved precision score. A CMFD method based on Binary Robust Invariant Scalable Key-points and image-blobs and proposed by P. Niyishanka et al. [13] proved to be effective against various post-processing attacks. Discrete Cosine Transform based detection of alteration in image was proposed by E.A. Vega et al. [14]. J.L. Zhong et al. [15] developed a dense moment feature index-based approach for detecting copy-move forgery in videos at the frame and pixel level. A. Roy et al. [16] proposed RLBP and g2NN based approach for addressing the problem of copy-move forgery for an identical but genuine object in an image. X. Tian et al. [18] developed a novel ORB feature extraction based novel similarity metric for effective and fast forgery detection.

4. Comparative Evaluation

Based on the survey done on the recent techniques/methods implemented by the researchers the comparative evaluation of the previous state of arts along with its benefits and limitations is provided in Table 1.

Table 1. Comparative evaluation of existing techniques for CMFD

Techniques	Benefits	Limitations
Two-pass hashing feature illustration and search approach [1]	-Resistant against geometric attacks and post processing attacks. -Improved computational efficiency deprived of iterations.	-CMFD results can be further improved, one way is by fusing Deep Neural Network.
Fast and effective CMFD method through adaptive keypoint extraction and processing [2]	-False matched pairs are removed. -Duplicated regions are localized.	-The source and target areas are not distinguished in the matching pair.
Blind CMFD through SVD and KS test [8]	-Forgeries can be detected at pixel level -High precision rate is achieved. -Reduced false positive rate lesser than 4%. -Proved robust for post-processing techniques.	-Larger computational time.

CMFD approach using tetrolet transform [11]	-Much accurate in identifying the size and shape of the forged region. -Effective results under geometric transformations and post processing.	-Cannot detect forged regions in noisy and distorted images.
Adaptive Attention and Residual Refinement Network for Copy-Move Forgery Detection [12]	-At pixel level AR-Net identifies the forged and the respective actual region. -Prediction mask is refined due to the use of AR-Net.	-Only single mode information is used in AR-Net.
Rotational CMFD methodology via region growing strategies and SIFT [3]	-Identifies the orientation difference among various copy-move regions.	-Matched key points and region growing method can be further improved.
Geometric transformation invariant CMFD via hybrid local features [4]	-Faster matching due to the use of SURF descriptor. -Improved results for geometric transformation.	-Multiple forgeries cannot be identified precisely. -Cannot identify forgery in highly similar regions.
CMFD for image forensics through super-pixel segmentation and Helmert transformation [5]	-Cost computation is less. -Tampered regions are more precisely located due to the use of SLIC algorithm.	-Forged region is not immune to smooth, recurring, and symmetric patterns
Probabilistic outline for copy-move forgery detection centred on Markov Random Field [6]	-Improved precision. -Energy minimization is undertaken using iterative method.	-Over segmentation due to SLIC.

5. Datasets

Due to the continuous advancements going on in the field of image forensic research a need of benchmark datasets is always there. A collection of both realistic and tampered images is provided in it which will help in finding more précised results for different techniques. A comprehensive overview of few publicly available datasets can be found in Table 2. which will help the researchers to decide which dataset to choose for a given CMFD approach.

Table 2. Publicly Available datasets

Dataset	Image Size	Total Images	Description
IMD dataset [1][9]	Average size of 3000×2300.	48	The copy-move images are tampered from 48 high resolution PNG base images.

GRIP dataset [2][11]	Images with the size 768×1024 or 1024×768 pixels.	80	It contains 80 original images of size 768 × 1024 or 1024 × 768.
FAU dataset [2]	3000×2000 pixels	48	This dataset is comprised of 48 high-resolution source images.
MICC-F220 [6][7][13]	The resolution of the image lies between 722×480 to 800×600 pixels	220	In this dataset 110 images are original and 110 are manipulated.
CoMoFoD dataset [7][8] [11] [12] [3] [13]	Varies from 512x512 to 3000x2000.	200	Each image is manipulated using 25 different combinations of tampering.
MICC-F600 [4][6]	Varying from 800×533 to 3888×2592 pixels.	160	This dataset constructs the testing images by applying different manipulations on copy-move region.
MICC-F2000 [4]	2048 × 1536 images.	2000	Out of 2000 images 700 of them are tampered.

6. Conclusion

In recent years, copy-move forgery detection has been one of the dominant areas of research. This paper aims at exploring various CMFD techniques. A comparative study of recent techniques has been made providing the advantage of choosing one technique over another along with its limitations. Also, description of relevant datasets for copy-move forgery detection is being provided which will help the researchers in the selection of the relevant datasets for efficient results for a given CMFD technique.

References

[1] J. Zhong and C. Pun, "Two-pass hashing feature representation and searching method for copy-move forgery detection", Information Sciences, Vol. 512, pp. 675-692, 2019.

[2] X. Wang et al., "A fast and high accurate image copy-move forgery detection Approach", Multidimensional Systems and Signal Processing, Vol. 31, pp. 857–883, 2019.

[3] Mahrishi M., Morwal S. (2020) "Index Point Detection and Semantic Indexing of Videos—A Comparative Review". In: Pant M., Kumar Sharma T., Arya R., Sahana B., Zolfagharinia H. (eds) Soft Computing: Theories and Applications. Advances in Intelligent Systems and Computing, vol 1154. Springer, Singapore.

[4] B. Soni, P. K. Das and D.M. Thounaojam, "Geometric transformation invariant block-based copy-move forgery detection using fast and efficient hybrid local features", Journal of Information Security and Applications, Vol. 45, pp. 44-51, 2019.

[5] H. Huang and A.J. Ciou, "Copy-move forgery detection for image forensics using the superpixel segmentation and the Helmert transformation", EURASIP Journal on Image and Video Processing, Vol. 68, 2019.

[6] B. Elhaminia, A. Harati and A. Taherinia, "A probabilistic framework for copy-move forgery detection based on Markov Random Field", Multimedia Tools and Applications, Vol. 78, pp. 25591–25609, 2019.

[7] P. Niyishaka and C. Bhagvati, "Copy-move forgery detection using image blobs and BRISK feature", Multimedia Tools and Applications, Vol. 79, pp. 26045–26059, 2020.

[8] B. Ahmed, T. A. Gulliver and S. Zahir, "Blind copy-move forgery detection using SVD and KS test", SN Applied Sciences, Springer, 2020.

[9] K.B. Meena and V. Tyagi, "A hybrid copy-move image forgery detection technique based on Fourier-Mellin and scale invariant feature Transforms", Multimedia Tools and Applications, Vol. 79, pp. 8197–8212, 2020.

[10] G. Gani and F. Qadir, "A robust copy-move forgery detection technique based on discrete cosine transform and cellular automata", Journal of Information Security and Applications, Vol. 54, 2020.

[11] K.B. Meena and V. Tyagi, "A copy-move image forgery detection technique based on tetrolet transform", Journal of Information Security and Applications, Vol. 52, 2020

[12] Y. Zhu, C. Chen, G. Yan, Y. Guo and Y. Dong, "AR-Net: Adaptive Attention and Residual Refinement Network for Copy-Move Forgery Detection", IEEE Transactions on Industrial Informatics, Vol. 16, pp. 6714 – 6723, 2020.

[13] P. Niyishaka and C. Bhagvati "Copy-move forgery detection using image blobs and BRISK feature", Multimedia Tools and Applications, Vol. 79, pp. 26045–26059, 2020.

[14] E.A. Vega, E.G. Fernandez, A.L.S. Orozco and L.J.G. Villalba "Copy-move forgery detection technique based on discrete cosine transform blocks features" Neural Computing &Applications, 2020.

[15] J.L. Zhong, C. M. Pun and Y.F. Gan, "Dense moment feature index and best match algorithms for video copy-move forgery detection", Information Sciences, Vol. 537, pp. 184-202, 2020.

[16] A. Roy, R. Dixit, R. Naskar and R.S. Chakraborty, "Copy-Move Forgery Detection with Similar But Genuine Objects", Vol. 755, pp. 65-77, 2020.

[17] K. Sunitha, A. N. Krishna, "Efficient Keypoint based Copy Move Forgery Detection Method using Hybrid Feature Extraction", International Conference on Innovative Mechanisms for Industry Applications, pp. 670-675, 2020.

[18] X. Tian, G. Zhou and M.Xau, "Image copy-move forgery detection algorithm based on ORB and novel similarity metric", IET Image Processing, Vol.10, pp. 2092-2100, 2020.

IOP Conf. Series: Materials Science and Engineering 1131 (2021) 012015 doi:10.1088/1757-899X/1131/1/012015

A dataset to evaluate Hindi Word Embeddings

Vimal Kumar Soni†, Dinesh Gopalani‡ and M C Govil§

Department of Computer Science and Engineering

Malaviya National Institute of Technology, Jaipur, India

E-mail: †ruvimals@gmail.com, ‡dgopalani.cse@mnit.ac.in, §govilmc@gmail.com

Abstract. The current trend to solve different challenges of Natural Language Processing utilizes various online crawling methods to fetch the data and applying different shallow or deep learning methods to develop models for the respective tasks on this data. Word vectors generated using such methods are being applied for several NLP challenges and such vectors are being evaluated on word similarity task. Not only huge data is available but also multiple datasets are available for the English language to evaluate the performance of the developed models. However, the scenario is not the same for Indian languages specifically for Hindi. Focusing this challenge, we propose a dataset to check word similarity in Hindi. The construction process and afterwards annotation process are described in details. To construct this dataset, first, 353 word-pairs from the most popular English dataset are selected and translated. Their translations are verified by Hindi Experts. These word pairs are finally annotated independently by 11 native Hindi speakers. Multiple criteria have been set to select the annotators for this task. The final dataset has been evaluated on CBOW and Skip-gram models.

1. Introduction

The Natural Language Processing deals with text written in different scripts and there is a huge gap in the pace of research among various languages. Enormous data from different domains as well as several tools are available for the English Language. The reason behind this is that English has been associated with computers since starting. The Internationalization process helped English to reach to the outskirts in a bundled pack with computers. However, for other languages, the situation is completely different as most of the computers have default language as English. In the past, computers used to have Indian language characters as non-standard characters however the scenario changed with the availability of language interface packs for Indian Languages and standardization of characters of scripts for Indian Languages. After the birth of Social Networking in 2003, the explosion of information happened and raw data of various languages started appearing on the internet and grew rapidly. This resulted in the availability of huge raw data which caters the need of bulk data for unsupervised learning.

With the advancement in neural networks in the form of shallow and deep learning, such crawled data are being used for unsupervised training to learn the semantic relations among the vocabulary words. These semantic relations are represented in the form of word vectors which can be used for various NLP challenges [1]. To evaluate such word vectors, word similarity task is quite popular and has already been accepted as a benchmark. Word vectors are applied to ambiguity resolution [2], automatic machine translation [3], next-word prediction [4], opinion mining [5] and question generation & answering [6]. The quality of word-vectors plays an important role in such applications as these vectors represent the semantics of the word in the form of vector. Therefore, the evaluation of such vectors is quite essential. Word similarity task has been proved to evaluate such vectors efficiently. A dataset which contains human-annotated semantic similarity index between word-pairs is used as a standard reference. Based

Content from this work may be used under the terms of the Creative Commons Attribution 3.0 licence. Any further distribution of this work must maintain attribution to the author(s) and the title of the work, journal citation and DOI.
Published under licence by IOP Publishing Ltd

on this reference, between the semantic similarity index and the respective word vectors, a correlation is calculated. Several research works are in progress to take leverage of word embeddings for the English language and different datasets are also available to evaluate these word embeddings but not for Hindi language. Though Hindi is the fourth most spoken language in the world yet the number of available computational resources for this language is low. A number of attempts are in progress to generate word embeddings for the Hindi language which can be used to address various NLP challenges, though the unavailability of a dataset to evaluate those word vectors is still a big challenge. This challenge motivated us to create a word similarity dataset to evaluate the Hindi language word vectors. The main contribution of this paper is to develop and provide the word similarity dataset for the Hindi language which is verified at multiple levels by Hindi language teachers and validated by inter-annotator agreement.

This paper is organized as follows: Initially, a summary of work in this area is presented in the next section. The process of construction of dataset is mentioned in Section-III. Section-IV describes the models and tuning parameters used to evaluate the dataset. The section-V explains about the evaluation and the section-VI discusses the results. The last section presents open research gaps.

2. Related Work

The English language is considered as a rich language in terms of availability of corpus and tools related to basic NLP tasks. In the same way, multiple datasets are available for the evaluation of word vectors. RG-65 [7] is a classical dataset which was released in 1965 and has motivated researchers of other languages to create word similarity datasets. In the year 1991, MC-30 [8] was released to show an inverse linear relationship between similarity of meaning and the discriminability of contexts.

Released in 2001, WordSim-353 [9] is considered as state of the art and one of the foremost datasets for benchmarking to evaluate the word representation vectors. Yang et al explored the idea of verb similarity in 2006 [10]. Szumlanski proposed a new dataset Rel-122 based on relatedness in 2013[11]. Li et al extended the idea of word similarity to multiword similarity and presented MWE300 in 2013 [12]. In 2015, Hill et al. used different POS categories to develop Simlex-999 dataset [13]. Gerz et al further explored the categories and developed a dataset for verb similarity in 2016 [14]. Different versions of WordSim-353 for other language are also developed. Netisopakul has used WordSim-353 and SimLex-999 to create Datasets for Thai Language [15]. Reference [16] developed multiple datasets for Indian languages namely Urdu, Marathi and Telugu using WordSim-353 and RG-65. There are various ways to generate word embeddings [17], however, all of these can be evaluated by word similarity task.

3. Dataset Construction

The steps involved in the construction of this dataset are shown in Figure-1.

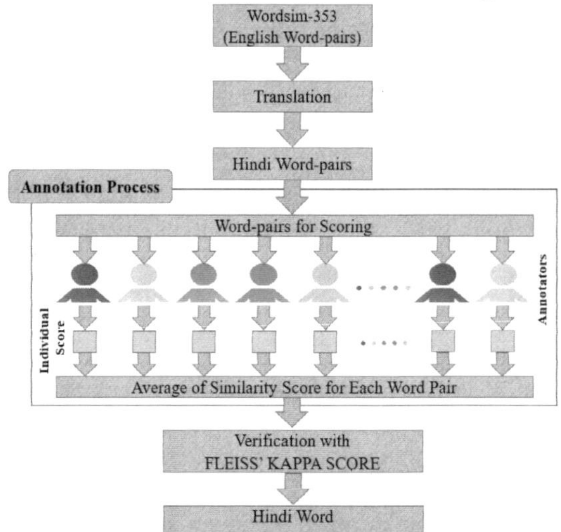

Figure 1. The process to Construct Hindi Word Similarity Dataset

3.1. Preparation of Data

Several datasets are available in the English language for word similarity task. To construct one for the Hindi language, we have taken the most popular word-similarity dataset from English which is WordSim-353. It contains 353 word-pairs which are domain-independent. This dataset has also been used to create datasets for other languages.

3.2. Translation

These word-pairs are then translated by an expert of Hindi language. The expert has been instructed to remain neutral for the translation of the individual word and not to take the context of any word from its pair. Few examples of Context-Free Vs Context-Based Translation from the word-sim353 are mentioned in the Table-1. Since the translation was based on the individual word so Context-Free Translation is preferred.

Table 1. Context-free vs context-based translation

Context-Free Translation				**Context-Based Translation**	
Word-1	***Word-1 Hindi Translation***	***Word-2***	***Word-2 Hindi Translation***	***Word Pair***	***Word-Pair Hindi Translation***
drug	दवा	abuse	गाली	drug-abuse	नशाखोरी
cell	कोशिका	phone	फ़ोन	cell phone	सैलफोन/ मोबाइल-फोन

There are words which are not part of Hindi Language, yet, they have become commonly accepted due to heavy usage, such words have been transliterated as it is instead of their exact translation. This has been done due to the fact that the exact translated word would have rare occurrence but the transliterated form has more occurrences. Few examples of such translation and transliteration cases are mentioned in Table-2 along with the accepted form used for the dataset.

Table 2. Translation vs Transliteration

Word	**Hindi Translation**	**Hindi Transliteration**	**Accepted Form**
computer	संगणक	कम्प्यूटर	कम्प्यूटर
report	रपट	रिपोर्ट	रिपोर्ट
theater	नाट्यशाला	थिएटर	थिएटर
film	चलचित्र	फिल्म	फिल्म

The specific names and acronyms have been accepted with their widespread transliterated versions in this dataset as shown in Table-3

Table 3. Transliteration of Proper Nouns

Word	**Hindi Transliteration**	**Accepted Form**
Arafat	अराफात	अराफात
Dollar	डॉलर	डॉलर
Jerusalem	यरूशलेम	यरूशलेम
OPEC	ओपीईसी	ओपीईसी

However, for the sake of consistency, only one meaning of a word has been taken irrespective of its occurrences in either side of word pair. Table-4 shows few words with multiple occurrences. The number of occurrences as the first part of pair and the last part of pair in the original wordsim353 dataset along with their most common Hindi Translation forms which are accepted for our dataset are also mentioned.

Table 4. Frequency of Occurrence

Word	Number of occurrences as the first part of pair	Number of occurrences as the last part of pair	Most Common Hindi Translation
money	12	2	धन
psychology	12	0	मनोविज्ञान
stock	7	2	शेयर
tiger	10	1	बाघ

The outcome of this step is a domain-independent set of 353 word-pairs in the Hindi Language.

3.3. Selection of Annotators

After finalizing the word-pairs in the Hindi Language, these are given to annotators to fill the similarity score manually for each word pair. There are three conditions kept to select an annotator for this task. First, Hindi must be the mother-tongue of the annotator and Second, he or she must have studied the Hindi language at least up to Senior Secondary Level and third, that he or she must have completed the graduation in any subject. Most of the annotators are either primary school teachers or postgraduate scholars and are directly related to the education domain. Total of 11 annotators provided this score.

3.4. Annotation and Scoring Process

The process of annotation is explained to the annotators. Annotators are asked to give a score for similarity to a word pair in a range of 0 to 10. Here 0 is to be assigned to totally non-similar word-pair while 10 is to be assigned to if both words in the word-pair are identical or have the perfectly same meaning. Annotators are also asked to fill the similarity of the words on their assessment instead of a mutual discussion with others to finalize the score. The experts involved in the translation process are not given the role of annotator to keep the annotation score free from any kind of bias which could occur due to the familiarity of word-pair during the translation process.

3.5. Post-processing & Finalizing the Similarity Score

To calculate the similarity score for each word-pair, the similarity scores provided by all annotators for the newly developed dataset are averaged over the collected 11 responses for each word-pair. The average score is taken as the final similarity score for a word-pair.

4. Model and Tuning Parameters

To evaluate the word vectors, first, such word embeddings are generated & trained using the following Corpus, Learning Models and Tuning parameters-

4.1. Corpus

The Monolingual corpus developed by IIT-Bombay is used to train the model [18]. This corpus contains approximately 45 million sentences. The newly developed dataset is used for testing purpose. A zero vector is assigned to Out-Of-Vocabulary (OOV) words. To ensure canonical Unicode representation, this monolingual corpus is normalized using the Indic NLP Library.

4.2. Learning Models

To learn the word embedding, the word2vec tool with CBOW & Skip-gram model is used. Along with these, the fastText Model is also used to generate the Hindi word vectors.

4.3. Tuning Parameters

As far as the tuning parameters are concerned, the dimension of the vectors is set to 300. The negative sampling is applied and set to 5 with a window of 5 words. The minimal number of word occurrences is also set to 5 to skip the rarely occurred words.

5. Evaluation

5.1. Inter Annotator Agreement

The semantics of a word not only changes according to context but also different readers can interpret that word in different ways. The impact of subjectivity can influence the annotation process. Therefore to check the impartiality and neutrality of the annotation process, an inter-annotator agreement is measured. This agreement defines the sanctity of the annotation along with the preciseness in the interpretation of one word with respect to the other word.

To calculate the inter-annotator agreement, several statistical measures are available like Cohen's kappa and Fleiss' kappa. Cohen's kappa is used to calculate the inter-annotator agreement between two annotators only whereas Fleiss' kappa is used as a measure to calculate the inter-annotator agreement for more than two annotators. For this dataset, the Fleiss' Kappa score for inter-annotator agreement is 0.79 which shows a good agreement among all annotators.

Table 5. Fleiss' Kappa Score

Dataset	Number of Annotators	Fleiss' Kappa score
Hindi	11	0.79

6. Result and Analysis

This word similarity dataset is evaluated on the state of the art models – Word2vec (with CBOW & Skipgram) and fastText as per the tuning parameters mentioned in section 4. The opted baseline scores are mentioned in the Table-6.

Table 6. Baseline Scores

System	Training Corpus Size	Score
Word2vec-CBOW	45 million Sentences	24.97
Word2vec-SkipGram	45 million Sentences	27.80
fastText	45 million Sentences	29.11

These trained models come across some words that do not appear during training (out of vocabulary words). It can be observed from Table 6 that fastText outperformed word2vec system. A random subset of the dataset is mentioned in the Table-7.

Table 7. Excerpts of HindiWordSim-353

S. No.	Word Pair		Similarity Score
	Word-1	Word-2	
1	बाघ	बाघ	10
2	बाघ	तेंदुआ	7.73
3	लड़का	बालक	9.36
4	टेलीफोन	संचार	5.55
5	महीना	होटल	2.36
6	राजा	पत्तागोभी	1.64
7	फुटबॉल	फुटबॉल	10
8	बैंक	धन	8.64
9	बिस्तर	अलमारी	4.09
10	टेलीविजन	रेडियो	6.73
11	सदी	राष्ट्र	3.18

The complete dataset is made available at https://github.com/ruvimals/Dataset-Hindi-WordSim353

7. Conclusion & Future Work

Multiple research works are being carried out using word vectors for natural languages. Resource-rich languages such as English or French are getting highly benefitted. However, widely spoken Hindi language is lacking due to scarcity of resources. Now with the current trend, many resources are being created and added to the resource-list of Indian languages. This work is a contribution towards the creation of resources for the Hindi language.

In future, this research work can be continued to develop the dataset with a ratio of POS categories. Another challenge is to create a domain-specific dataset to carry out the research work in domains like agriculture or tourism.

References

[1] Mikolov, Tomas, Kai Chen, Greg Corrado, and Jeffrey Dean. "Efficient estimation of word representations in vector space." arXiv preprint arXiv:1301.3781 (2013).

[2] Hadiwinoto, Christian, Hwee Tou Ng, and Wee Chung Gan. "Improved word sense disambiguation using pre-trained contextualized word representations." arXiv preprint arXiv:1910.00194 (2019).

[3] Choi, Heeyoul, Kyunghyun Cho, and Yoshua Bengio. "Context-dependent word representation for neural machine translation." Computer Speech & Language 45 (2017): 149-160.

[4] Tenney, Ian, et al. "What do you learn from context? probing for sentence structure in contextualized word representations." arXiv preprint arXiv:1905.06316 (2019).

[5] Rana, Ankur, Vishal Goyal, and Vimal K. Soni. "Renaissance of Opinion Mining." International Conference on Information Systems for Indian Languages. Springer, Berlin, Heidelberg, 2011.

[6] Medved, Marek, and Ales Horák. "Sentence and Word Embedding Employed in Open Question-Answering." ICAART (2). 2018.

[7] Rubenstein, Herbert, and John B. Goodenough. "Contextual correlates of synonymy." Communications of the ACM 8.10 (1965): 627-633.

[8] Miller, George A., and Walter G. Charles. "Contextual correlates of semantic similarity." Language and cognitive processes 6.1 (1991): 1-28.

[9] Finkelstein, Lev, et al. "Placing search in context: The concept revisited." Proceedings of the 10th international conference on World Wide Web. 2001.

[10] Mahrishi, M., Hiran, K. K., Meena, G., & Sharma, P. (2020). Machine Learning and Deep Learning in Real-Time Applications. IGI Global.

[11] Szumlanski, Sean, Fernando Gomez, and Valerie K. Sims. "A new set of norms for semantic relatedness measures." In Proceedings of the 51st Annual Meeting of the Association for Computational Linguistics (Volume 2: Short Papers), pp. 890-895. 2013.

[12] Li, Peipei, Haixun Wang, Kenny Q. Zhu, Zhongyuan Wang, and Xindong Wu. "Computing term similarity by large probabilistic isa knowledge." In Proceedings of the 22nd ACM international conference on Information & Knowledge Management, pp. 1401-1410. 2013.

[13] Hill, Felix, Roi Reichart, and Anna Korhonen. "Simlex-999: Evaluating semantic models with (genuine) similarity estimation." Computational Linguistics 41, no. 4 (2015): 665-695.

[14] Gerz, Daniela, Ivan Vulić, Felix Hill, Roi Reichart, and Anna Korhonen. "Simverb-3500: A large-scale evaluation set of verb similarity." arXiv preprint arXiv:1608.00869 (2016).

[15] P. Netisopakul, G. Wohlgenannt and A. Pulich, "Word Similarity Datasets for Thai: Construction and Evaluation," in IEEE Access, vol. 7, pp. 142907-142915, 2019, doi: 10.1109/ACCESS.2019.2944151.

[16] Akhtar, Syed Sarfaraz, et al. "Word similarity datasets for indian languages: Annotation and baseline systems." Proceedings of the 11th Linguistic Annotation Workshop. 2017.

[17] Soni, Vimal Kumar, Dinesh Gopalani, and M. C. Govil. "Resurgence of Deep Learning: Genesis of Word Embedding." Smart Innovations in Communication and Computational Sciences. Springer, Singapore, 2019. 129-139.

[18] Anoop Kunchukuttan, Pratik Mehta, Pushpak Bhattacharyya. "The IIT Bombay English-Hindi Parallel Corpus." Language Resources and Evaluation Conference. 2018.

Impact of Driving Factors on Cloud Computing Adoption in the Higher Education

Kamal Kant Hiran

Research Fellow, Aalborg University, Copenhagen, Denmark

kkh@es.aau.dk

Abstract. The aim of this research is to explore the sociocultural and economic factors involved in the adoption of emerging technology, i.e., the cloud at the University in Ethiopia. Organizations' growth and development depend on the various factors that influence productivity and efficiency to improve the work culture. The data collected from the relevant key actors or stakeholders from the higher education system i.e., University of Gondar, Ethiopia, and complied in a purposive way. The researcher identified the most inducing factors under the sociocultural categories are societal belief with regression weight SoCF_1=0.821, followed by ideologies or value of individual towards CC, language barrier, and awareness towards the cloud. Similarly, the most inducing economic factors were high investment cost with regression weight EcoF_1=1.096, followed by per capita income, government grants and infrastructure development. The framework from the full-scale analysis will theoretically lead to the creation of a discussion model that can be used to examine cloud adoption in higher education.

Keywords: sociocultural, economic factors, stakeholders, government grants

1. Introduction

The fast-moving technological age gives institutions unprecedented potential for improving their productivity (Ali & Osmanaj, 2020). By using the emerging technologies, the efficiency and services can be improved in the institutions. Cloud Computing has evolved as technology enables businesses to operate independently on demand and IT capabilities without costly upfront investments, regular software upgrades and staff retraining (Dadgar, Vithayathil, & Osiri, 2017).

As the technologies are growing day by day such as Cloud computing, Edge computing, Machine learning, Deep learning, Internet of Things, Intelligent Cloud, etc. which are widely adopted by individuals and institutions to perform their operations (Shi, 2017). For the effective adoption of any emerging technologies, several factors must be examined before its adoption. Out of several predictive factor's researcher focus in this study on sociocultural and economic factors in higher education (Yani Supriani, Giyanti, & Tb. Sofwan Hadi, 2020). The area of the study is the adoption of CC in higher education in the underdeveloped country of Ethiopia (Hanlon, Whitley, Wondimagegn, Alem, & Prince, 2009). There is a limited research on the application of technology in the developing economics, along with social and economic concepts. As has been reiterated, developing countries face various challenges in technological innovation. Ethiopia, like other developing countries, lags developed nations weak infrastructure, costly investment, lack of awareness and slow response to

Content from this work may be used under the terms of the Creative Commons Attribution 3.0 licence. Any further distribution of this work must maintain attribution to the author(s) and the title of the work, journal citation and DOI.
Published under licence by IOP Publishing Ltd

technocrats (Hiran & Henten, 2020). It is therefore necessary to develop a comprehensive scientific understanding of adoption predictors in different cognitive development and geographic contexts. The economic and social development change in many developed nations and the awareness may also help to shape the successful adoption of CC policies in Ethiopia (Chaiklin & Hedegaard, 2013).

The objective of the research is to find out the most prominent sociocultural and economic factors for the cloud adoption in the higher education at the University in Ethiopia?

2. Review of Literature

According to Vygotsky (Sawyer & Stetsenko, 2018), at the beginning of the period, a major, rare, and highly unique relationship develops between the youngster and the realization. Vygotsky explained, "The social development measure shows the starting point in time for all complex changes occurring during the period in question". It decides absolutely the forms and the path the social becomes human. (Barrios, Guilera, Nuño, & Gómez-Benito, 2020) carried out a Delphi analysis of firm-level issues with certain IT specialists representing clients, suppliers, and scholars across a wide range of nations, sectors, and institutions, and found that economic construction was vital. Several other studies also indicated social, economic, customer behavior and regulatory factors were the obstructions. (Barrios et al.,2020) emphasizes that the positioning of social rather than human mechanisms is a primary element of socio-cultural theory in developing higher mental functions. It is correct that intellectual growth could not be seen in a genetically mediated pattern, but in experiences in a social and cultural context. Learning is therefore seen as going to lead or promoting academic success. (Lev Vygotsky, 1896) pointed to the solution of 2 (two) forms of biological reductionism, the natural maturation of the physiological and societal brain, the natural learning of the cultural assets of the language, community, etc. Roosevelt (Yani Supriani et al., 2020) argued that the primary goal of Vygotskian learning is to keep students in their ZPD (Zone of Proximal Development) as long as possible by presenting students with encouraging and culturally relevant teaching and solution activities. (Gajdamaschko, 2015) point to the emergence of skills and processes not inherent, stable, and observable in Vygotsky psychology, but rather because of the experience of social experiences of a person around the world.

2.1. Zone of Proximal Development (ZPD)

Vygotsky (Lev Vygotsky, 1896) introduced the idea of ZPD (see figure 1) and described it as the remoteness between the actual level of development centered on Independent problem-solving and threshold of potential growth estimated by problems addressed under supervision and in cooperation with adequate peers in society; (Fani & Ghaemi, 2011) 'the growth of better psychological factors.' That is, Vygotsky interpreted the ZPDs as identifying the existing and real learners' level growth as well as the next stage reached by using semiotic and environmental mediating methods and encouraging the use of adults or peers. The theory is that people learn better by working together with other people in shared cooperation and by working with professional people, they learn and internalize new ideas, psychological techniques, and skills. (Chaiklin & Hedegaard, 2013) believes it is necessary to note that the interest of Vygotsky is to establish a theoretical framework for adequate pedagogical procedures, including concepts of the potential educational gathering of students and the identification of interventions of individual learners, as an initial step towards understanding how he developed the field of proximal growth. His theory (Lev Vygotsky, 1896) suggested the below important elements: a. The fundamental significance of heritage and traditions as a mechanism of growth, b. Origin of internally and externally resources that shape the structure of mental function, c. comprehensive ZPD.

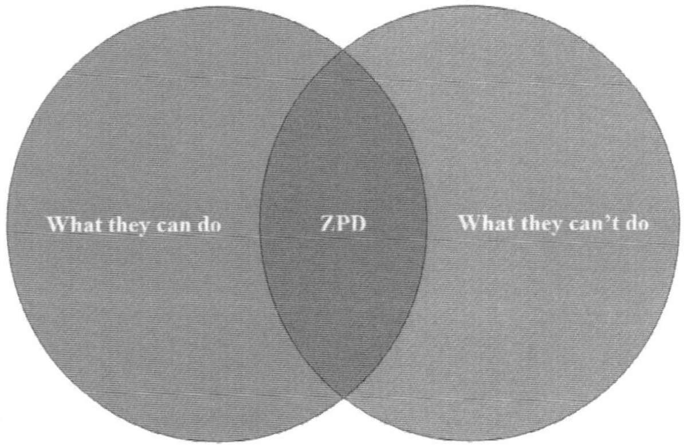

Figure 1: Zone of Proximal Development

3. Theoretical framework

This section describes the conceptual model structure for the analysis. Since cloud adoption is the subject of this study, so the researcher outlined Vygotsky's ideas (Lev Vygotsky, 1896) that articulate the central role of social interaction in the creation of understanding, as he firmly believes that culture plays a key role in making sense. Therefore, the researcher reviewed the societal and economic parameters at the outset then examined research related to this study. Below is the proposed theoretical framework based on the sociocultural and economic factors and cloud adoption in higher education (see Figure 2).

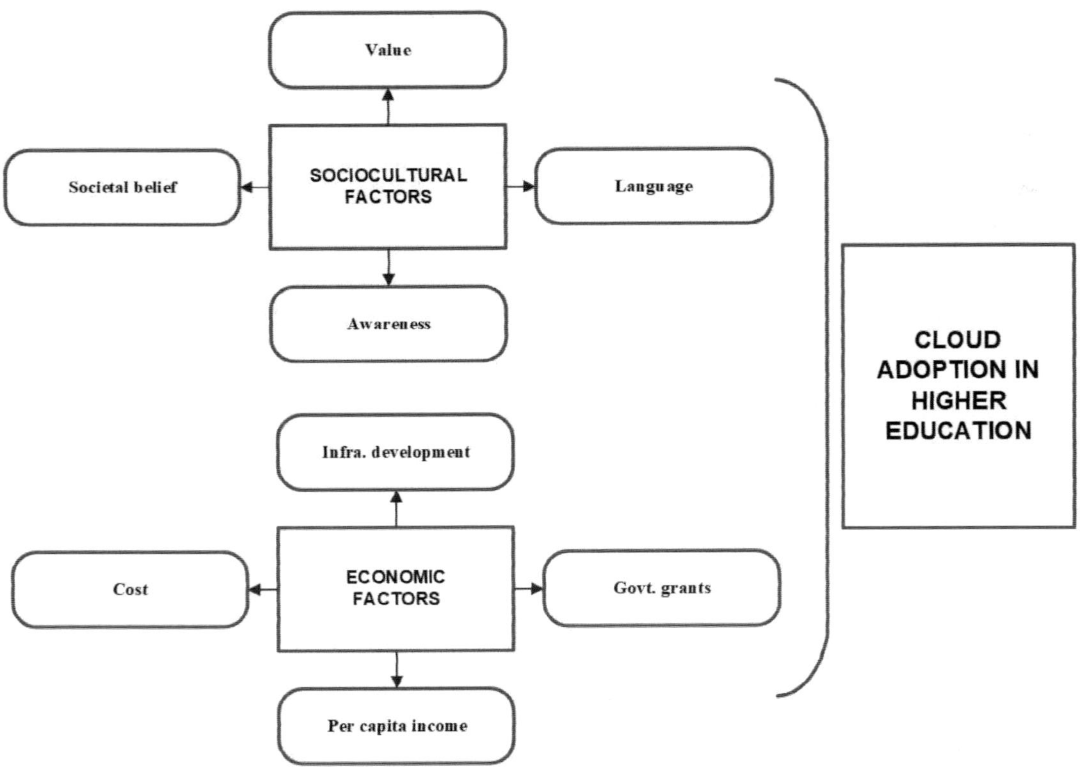

Figure 2: Theoretical framework

IOP Conf. Series: Materials Science and Engineering 1131 (2021) 012016 doi:10.1088/1757-899X/1131/1/012016

4. Research Methodology

The persistence of the study is analyzing frequently reported sociocultural and economic factors to widespread adoption of cloud technology. This goal has been followed by examining the selected item that has a profound effect on CCA.

Sample Size: To go for an extensive study UoG in Amhara region, Ethiopia was selected. 125 respondents were targeted by the report, but only 95 of them tried all the declaration and thus the research was limited to those defendants. There are two parts of the questionnaire. The first portion consists of 4 (four) questions on the demographic details of respondents (see Table 1). The second portion is made up of 10 (ten) statements based on sociocultural-economic constructs, to study a comprehensive view concerning the adoption of CC. For analyzing each statement, the opinion of the selected actors has been elicited on Likert five theme.

Collection of Data: Both primary and secondary data have complied in this study. Main data obtained through a survey form and through interviews with different groups of sample survey respondents. The secondary data collected from the published records of online companies, journals, books, etc.

Tools of Analysis: The statistical paraphernalia used for the study included Reliability test, Regression, KMO test, etc.

5. Data Analysis and Interpretation

Table 1: Demographic Description

Factor	Classification	Frequency	Percent	Cum Percent
Gender	Male	55	57.90	57.90
	Female	40	42.10	42.10
		95		100
Age	Below 30	48	50.50	50.50
	31-40	29	30.60	80.60
	Above 41	18	18.90	100
		95		
Education	UG	35	36.80	36.80
	P.G	50	52.60	89.40
	Others	10	10.60	100
		95		
Work Exp	< 1 year	25	26.30	26.30
	1-5 years	52	54.80	81.10
	> 5 years	18	18.90	100
		95		

The above table 1 articulates that the sample consisted of 57.90 percent male and 42.10 percent female, predominantly 48 respondents were below the age group of 30 years, 29 respondents were between 31-40 years and only 18 persons were above the age of 41. Further, most of the respondents were post-graduate 50, under-graduate 35, and others were only 10 respondents. 26.30 percent of participants having work experience less than one year whereas most of the participants i.e., 52 were under the category of 1-5 years, and eventually on 18 participants having experience of more than 5 years.

IOP Conf. Series: Materials Science and Engineering 1131 (2021) 012016 doi:10.1088/1757-899X/1131/1/012016

Primarily, to test the reliability of these items, Cronbach's Alpha score was computed that revealed a score of 0.785 that, outlined that the statements were reliable enough for further statistical investigation.

Table 2: Multiple Regression Results of Sociocultural Factors

(N=95)

Variables	Codes of items	Ad. R2	B	ANOVA	Sign.
Sociocultural Factors	SoCF_1	0.643	0.821	34.840	$.000^b$
	SoCF_2		0.223		
	SoCF_3		0.217		
	SoCF_4		0.226		

Dep. Variable: CCA

The above table 2 depicted the result of regression tool and the most influencing factors were societal belief (SoCF_1), b=0.821, language barrier (SoCF_2), b=0.223, awareness towards the cloud (SoCF_3), b=0.217, and ideologies or value of individual towards CC (SoCF_4), b=0.226. The f-value=34.840 of selected constructs at less than 0.05 level clearly enunciates that all these five factors influence and explain (adjusted R^2 =0.64) the variance of approximate 64 percent users' intention toward cloud adoption.

Table 3: Multiple Regression Results of Economic Factors

(N=95)

Variables	Codes of items	Ad. R2	B	ANOVA	Sign.
Economic Factors	EcoF_1	0.863	1.096	119.38	$.000^b$
	EcoF_2		0.033		
	EcoF_3		0.531		
	EcoF_4		0.749		

Dep. Variable: CCA

The above table 3 articulates the outcome of regression and the most inducing factors were high investment cost (EcoF_1), b=1.096, Infrastructure development (EcoF_2), b=0.033, government grant & subsidies (EcoF_3), b=0.226 and per capita income (EcoF_4), b=0.749. The f-value=119.38 of selected builds at less than 0.05 level clearly express that all these five factors sway and illuminate (adjusted R2 =0.64) the variance of 86 percent with the users' intention toward cloud adoption. Thus, it can be inferred that cooperative activities among the selected variables are required to foster the CCA.

6. Discussion and Conclusion

The rapidly evolving digital revolution provides businesses unparalleled opportunities for increased results through operating productivity, improved customer support, growth of the market, and innovation. It is necessary to get acquainted with new-age technology for the creation of an organizational culture of learning, development, and foster creativity. The creation of technology standards includes the involvement of members from various organizations in the CCA. It is advised that users from all organizational members be actively involved in the strategic or creation process right from the outset, instead of creating "top-down" methods. In order, to increase awareness and the pace of adoption of the cloud, it is important to provide opportunities for early users who are unable to enjoy the full advantages of technological innovation. Since economic and socio-cultural challenges cannot be resolved by policymakers alone, it is important to provide a detailed overview of the main factors that affect the adoption of the HEI cloud, and the same has been examined. Thus, the study examines the role of adopted socio-culture and economic aspects as an antecedent of cloud technology adoption. The results of the complete assessment can contribute to developing a knowledge model that could be used to examine. It will also contribute to scholarly literature through empirical evidence from study results and include future economic and sociocultural adoption. The result of sociocultural regression depicts that beta value is 0.821 for societal belief, language barrier 0.223, awareness towards cloud 0.217, ideologies or value of individual towards cloud is 0.226. Similarly, the most inducing economic factors were high regression weight via. 1.096 for investment cost, infrastructure development 0.033, government grant and subsidies 0.226 followed by per capita income 0.749. The f-value of selected constructs were significant and answered the research objective (p-value less than 0.05) that clearly enunciate that all these factors not only influence CCA but also explain users' intention toward cloud adoption.

The findings of the proposed study would also include useful knowledge backed by empirical evidence that can be used in academics designing and promoting CC projects. Finally, the analysis will contribute to future research on best technology implementation models applicable to emerging-country contexts. Further, to address the social and economic obstacles associated with configuration of cloud; therefore, the university should also use key members that actively foster change and encourage the transfer of information and the dissemination of knowledge across the cloud.

References

[1] Ali, O., & Osmanaj, V. (2020). The role of government regulations in the adoption of cloud computing: A case study of local government. Computer Law and Security Review, 36. https://doi.org/10.1016/j.clsr.2020.105396

[2] Barrios, M., Guilera, G., Nuño, L., & Gómez-Benito, J. (2020). Consensus in the delphi method: What makes a decision change? Technological Forecasting and Social Change, (November 2019), 120484. https://doi.org/10.1016/j.techfore.2020.120484

[3] Chaiklin, S., & Hedegaard, M. (2013). Cultural-Historical Theory and Educational Practice : Some Radical-Local Considerations Teoria Histórico-Cultural E Práticas Educacionais : Algumas Considerações Radicais. 30–44.

[4] Kant Hiran, K., Henten, A., Shrivas, M. K., & Doshi, R. (2018). Hybrid EduCloud Model in Higher Education: The case of Sub-Saharan Africa, Ethiopia. In 7th International Conference on Adaptive Science & Technology (ICAST) [8507113] IEEE.

[5] Fani, T., & Ghaemi, F. (2011). Implications of Vygotsky's zone of proximal development (ZPD) in teacher education: ZPTD and self-scaffolding. Procedia - Social and Behavioral Sciences, 29(Iceepsy), 1549–1554. https://doi.org/10.1016/j.sbspro.2011.11.396

[6] Gajdamaschko, N. (2015). Vygotsky's Sociocultural Theory. International Encyclopedia of the Social & Behavioral Sciences: Second Edition, 25, 329–334. https://doi.org/10.1016/B978-0-08-097086-8.23203-0

[7] Mahrishi M., Hiran K.K., Doshi R. (2020) "Selection of Cloud Service Provider Based on Sampled Non-functional Attribute Set". In: Abraham A., Siarry P., Ma K., Kaklauskas A. (eds) Intelligent Systems Design and Applications. ISDA 2019. Advances in Intelligent Systems and Computing, vol 1181. Springer, Cham.

[8] Hiran, K. K., & Henten, A. (2020). An integrated TOE–DoI framework for cloud computing adoption in the higher education sector: case study of Sub-Saharan Africa, Ethiopia. International Journal of Systems Assurance Engineering and Management, 11(2), 441–449. https://doi.org/10.1007/s13198-019-00872-z

[9] Lev Vygotsky. (1896). Lev Vygotsky (1896–1934) - Cultural-Historical Theory, Education and Cognitive Development - Thinking, Psychology, Functions, and Self - StateUniversity.com. Retrieved November 24, 2020, from https://education.stateuniversity.com/pages/2539/Vygotsky-Lev-1896-1934.html

[10] Sawyer, J. E., & Stetsenko, A. (2018). Revisiting Marx and problematizing Vygotsky: a transformative approach to language and speech internalization. Language Sciences, 70, 143–154. https://doi.org/10.1016/j.langsci.2018.05.003

[11] Senyo, P. K., Effah, J., & Addae, E. (2016). Preliminary insight into cloud computing adoption

in a developing country. Journal of Enterprise Information Management, 29(4), 505–524. https://doi.org/10.1108/JEIM-09-2014-0094
[12] Shi, L. (2017). The application of cloud computing technology in network data security. Agro Food Industry Hi-Tech, 28(1), 912–916. https://doi.org/10.1007/978-3-319-01766-2_99
[13] Yani Supriani, Giyanti, & Tb. Sofwan Hadi. (2020). Conjecturing Ability Dalam Pembelajaran Daring Masa Pandemi Covid-19. Inomatika, 2(2), 69–77. https://doi.org/10.35438/inomatika.v2i2.201

List of Acronyms Used

- CC – Cloud Computing
- CCA – Cloud Computing Adoption
- HEI – Higher Education Institution
- SoCF – Sociocultural Factor
- EcoF – Economic Factor
- ZPD – Zone of Proximal Development
- KMO – Kaiser-Meyer-Olkin

Automation of discussion board evaluation through keyword extraction techniques: a comparative study

Gomathi Thiyagarajan [1 3], Prasanna S [2] and Uma B [3]

[1]Research Scholar, School of Computing Science, VISTAS, Chennai, India
[2]Professor & Head, Department of Computer Applications, VISTAS, Chennai, India
[3]Department of Computer Applications, CMR Institute of Technology, Bengaluru, India
gomathi.t@cmrit.ac.in

Abstract. With the current trend in lieu of the ongoing pandemic, online learning and technology-based pedagogy receive worldwide attention. There are several pedagogical practices one can use to evaluate students mindful and the method will vary based on the learning objective. In this paper, our goal is to dive deeper into identifying the best keyword extraction technique which can be closely mapped with human evaluation. We report the results of three popular keyword extraction techniques specifically RAKE, TF-IDF, and Semantic Fingerprinting on the dataset generated from the discussion board post of a Learning Management System. The results illustrate that the TF-IDF algorithm shows the highest correlation with a human evaluation with 0.76 correlations, RAKE with 0.36, and Semantic Fingerprinting with 0.58. It was also identified Semantic Fingerprinting had a lowest mean square error of 18.26.

Keywords: Automatic Keyword Extraction, Discussion Board evaluation, RAKE, TF-IDF, Semantic Fingerprinting

1. Introduction

Educational Institutions are investigating new pedagogical approaches to figure out and handle difficulties such as average class size, facilitate dynamic, active cooperation among different companion of students [1]. Assessment is an essential component for viable learning. As Vonderwell et al. [2] indicated, web-based learning contexts include distinct characteristics when contrasted with face-to-face settings because of the asynchronous collaboration among the teacher and learner. A well-organized strategy [3] and assessment components would be required in place as meaningful collaboration in both online and blended settings are incorporated by effective learning communities.

The discussion board has become the most popular model of interaction to create a virtual community of inquiry, allowing peers to comment, question, and respond. It is available as both stand-alone and built-in tools in Learning Management Systems (LMSs) typically allows multi-threaded discussion and records every post. Moore, R. L et al. [4] endowed that few segments of discussion colloquy backed the advancement of higher-order thinking; they further suggested that the type of doled out assignment and the forum elicit the type of higher-order thinking in a networked conversation. For decades, educationalists have done their research on automating the evaluation process.

Martin F et al. [5] stated the recommendations from faculties for assessment includes traditional, authentic, rubrics, interviews, surveys, systematic learning analytics, and peer reviews. Amongst the various evaluation mechanisms, this study primarily focuses on keyword-based evaluation. In this paper,

Content from this work may be used under the terms of the Creative Commons Attribution 3.0 licence. Any further distribution of this work must maintain attribution to the author(s) and the title of the work, journal citation and DOI.
Published under licence by IOP Publishing Ltd

our goal is to explore the effective keyword extraction technique that can be used to automate the assessment of the posts in Learning Management Software. Keywords from individual posts were extracted using various techniques and compared with Lecturer given keywords to identify the percentage of matches between the extracted and given keywords. The posts were also manually evaluated and this correlation between the two was presented in this work.

The rest of this paper contains; Keyword Extraction techniques in Section 2 covers the popular technique used to excerpt keywords and discuss the popular algorithms that are being employed to pull out the keywords to automate evaluation. Section 3 and 4 briefly explains the data collection and evaluation method used in the current research and their results. Finally, Section 5 furnishes the conclusion of this study.

2. Keyword Extraction Techniques

Keywords are the small scale units that can sum up the content of a document. Keywords are regularly used for many reasons like recovering records during a web search or summing up reports for indexing and they are likewise used to pin down the most applicable information from the context. Automatic keyword extraction methods are generally used in Text Mining, Natural Language Processing, and data recovery to represent large sets of words to a small bunch of words.

Beliga, S et al. [6] stated different techniques that can be used for extracting keywords are based on statistics, linguistic, graph-based, and machine learning approaches. The table below shows the techniques and various approaches used for keyword extraction. For the current research, we used the RAKE, TF-IDF, and Semantic Fingerprinting technique to extract the keywords from the text, and the flow diagram of the techniques is shown in figure 2.

2.1. RAKE

Rapid Automatic Keyword Extraction (RAKE) [7] [8] is an unsupervised, language, and domain-independent technique for extracting keywords. The RAKE technique depicted below Figure automates the keyword build process quickly and expressively through the following steps.

- A document comprises multiple phrases and sentences from which keywords have to be extracted.
- The input parameter for RAKE consists of stop words, phrase delimiters, and word delimiters.
- RAKE through these input parameters partitions the document into candidate keys.
- The document text is split into an array of words called a candidate key.
- Every candidate key is given a score (degree of frequency) by analyzing the co-occurrences graph. **The degree of frequency = Degree of word / Degree of frequency**
- The keywords with top T-Scores are selected as keywords for the document.

2.2. TF-IDF

The Term Frequency-Inverse Document Frequency [9] [10] is a methodology used to discover the significance of sentences involving words. It registers values for each word in a record through an opposite extent of the recurrence of the word in a particular report to the level of archives the word appears in. The equation used to process tf-idf is given in (1).

$$\text{tf-idf}_{(t,\,d)} = \text{tf}_{(t,\,d)} * \text{idf}_{(t)} \qquad (1)$$

The formula that is used to compute the IDF is computed as in (2)

$$\text{idf}_{(t)} = \log\,[n\,/\,(\text{df}_{(t)} + 1)]) \qquad (2)$$

Table 1. Techniques and approaches to Keyword Extraction

Approaches	Techniques
Simple Statistic Approach	Word Frequency, Word Co-Occurrences, TF-IDF, RAKE
Linguistic Approach	Discourse Markers, Semantic information, Linguistic information
Graph-Based Approach	Word Co-Occurrences, Labelled Edges, Neighbourhood Size
Machine Learning Approaches	Support Vector Machine, Deep Learning, Conditional Random Fields

The word with a high TF-IDF score has a solid relationship with the report in which it shows up. The term frequency refers to the heaviness of a term that occurs in a report. Document Frequency is the quantity of archives in which the word is available.

TF - denotes the occurrences of a specific term t in document d
IDF - denotes the no. of documents in the collection containing the term t.

2.3. Semantic Fingerprinting

As per ongoing discoveries in neuroscience, the brain utilizes Sparse Distributed Representations (SDR) to handle data and information. Sparse implies neurons are dynamic simultaneously, contrasts to the common dense and distributed represents the significance of the pattern in it. The concept driving Cortical.io's Retina [11] comprises changing over the symbolic representation of language into an SDR structure with the goal that it turns out to be mathematically computable. Language is first decomposed into words. The terms are then changed over into SDRs by the Retina Engine. Each word is described by a heap of 16,000 semantic features, which catch the lexical semantics of a word in its normal setting. Two words become comparable in their SDR structure if they are reasonably related. Semantic similarity can be computed using a simple measure like the Euclidean Distance. Figure 1 represents the semantic folding model used in the Semantic Fingerprinting approach.

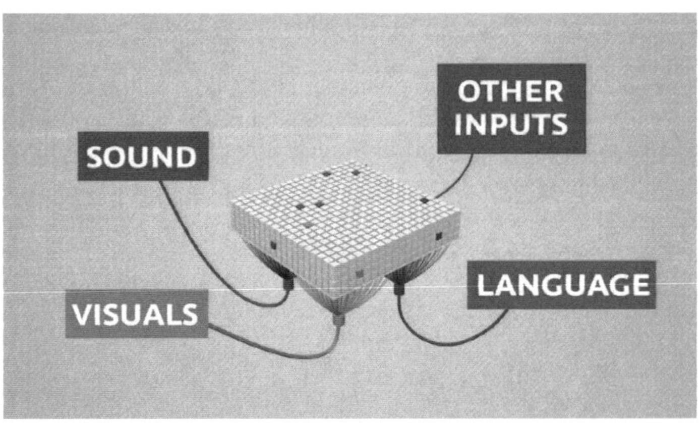

Figure 1. Semantic Folding: a new model for natural language understanding [digital image]. Retrieved from https://www.cortical.io /

The semantic fingerprinting [12] approach decides the significance of each term in the content by producing a semantic fingerprint for each term with the finger print of text determining the overlap looking at the frequency of the term in content with the frequency of the term in the whole retina semantic space.

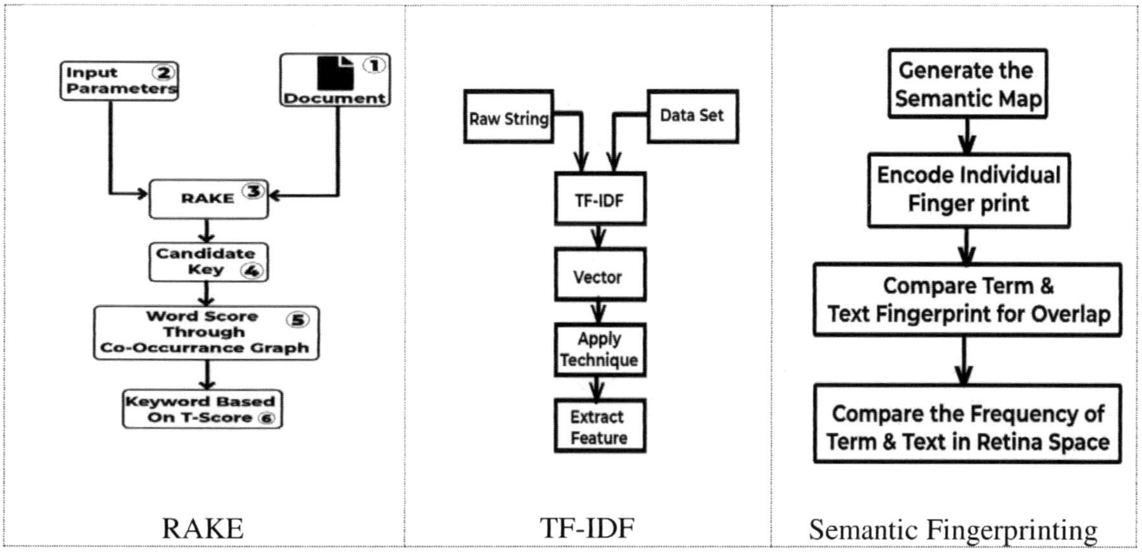

Figure 2. Flow Diagram of RAKE/TF-IDF and Semantic Fingerprinting technique

3. Data evaluation technique

The dataset used in the current research consists of 52 responses to a post by the PG (Post Graduate) students on a topic discussed in the course Big Data Analytics. Students were asked to post their understanding of the topic covered in the online class simultaneously through Google Classroom. All forum posts have been first extracted from Google Classroom through the script written in PHP. The name of the student, timestamp, and the posted responses were extracted and placed in a spreadsheet to apply keyword extraction algorithms for instant evaluation. Figure 3 illustrates the data collection method and evaluation strategy carried out by the researchers during this work.

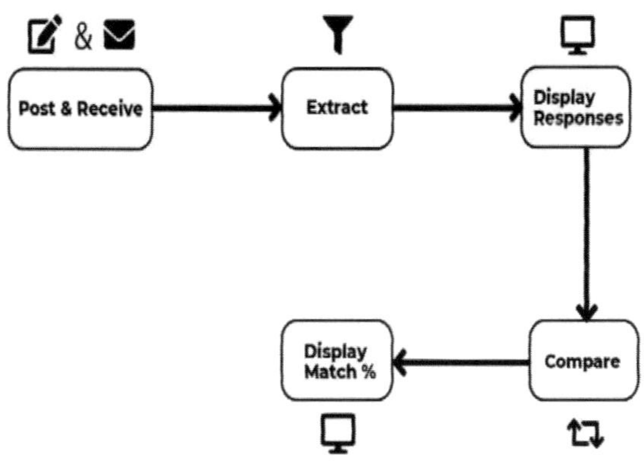

Figure 3. Data collection and Extraction

Figure 4. Word Cloud of posts

IOP Conf. Series: Materials Science and Engineering 1131 (2021) 012017 doi:10.1088/1757-899X/1131/1/012017

- Post and Receive mainly consists of posting a question in a Learning Management System and receiving responses from learners in a stipulated time.
- Extract here performs extraction of student response from the discussion board (LMS used in current research is Google Classroom). The Word Cloud is shown in figure 4 shows the graphical representation of word frequency that appears in the forum posts by students.
- Extraction also represents extracting keywords using keyword extraction techniques or algorithms. Table 2 displays the detection of keywords using various keyword extraction techniques for a sample post.
- Compare consists of logic to compare the extracted keyword with keywords suggested by experts and its synonym using wordnet.
- Display match % displays the matches between keywords percentage.

Table 2. Sample keyword extracted

RAKE	TF_IDF	Semantic Fingerprinting
max value	Reduce	value
key-value pair	values	max
type of output	particular	values
Value	temperature	phase
particular year	year	output
reduce phase	reducer	pairs
highest recorded temperature	types	temperature
type of input	Output	types
intwritable type	Key	key
Text	pairs	variables

4. Results and discussion

This study examined the keyword extraction algorithm to perform an instant evaluation of discussion board posts during an online class. In general, it is difficult to set up the best technique for keyword extraction since certain algorithms work better in specific circumstances. In this section, we evaluated three keyword extractions methods namely RAKE, TF-IDF, and Semantic fingerprinting methods with the human evaluation of the discussion board post. The results of correlation between Human, RAKE, TF-IDF, and Semantic fingerprinting is represented in figure 5. TF-IDF outperforms the other approaches by returning relevant keywords to a particular query and shows the nearest correlation with human evaluation. RAKE performs the weakest, due to optimizing long keyword for accurate phrases. The third method semantic fingerprinting performs better than RAKE and it represents text as a semantic fingerprint by using a Boolean vector with the lowest error rate. Figure 6 shows the Pearson's r correlation between Human evaluation and RAKE/TF-IDF/Semantic Fingerprinting evaluation.

Table 3. Comparison between various techniques

Source	Correlation Coefficient	Mean Square Error	Root Mean Square Error	Mean Absolute Error	Coefficient of Determination
RAKE	0.358	43.082	6.564	5.888	0.082
TF-IDF	0.763	32.354	5.688	4.768	0.536
Semantic Finger Printing	0.577	18.257	4.273	3.419	0.261

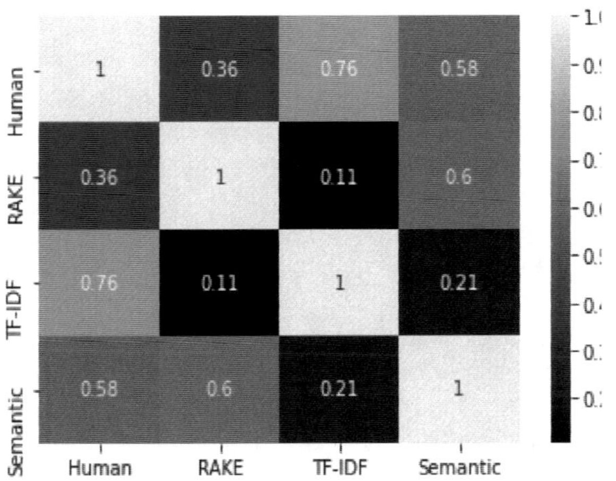

Figure 5. Correlation Matrix

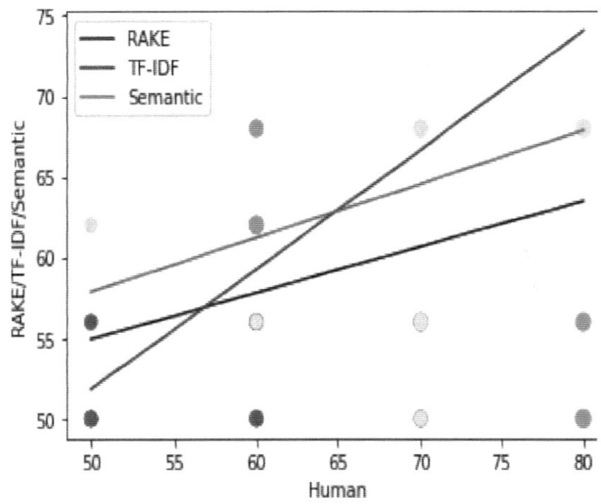

Figure 6. Correlation plot

5. Conclusion

This study primarily focused on identifying the keyword extraction technique which can be used to instantly evaluate the discussion board post of students based on their understanding of the topic covered in a session through online mode. This can be an effective pedagogical evaluation for the faculties to quickly get to know the understanding of the just-concluded session and to effectively plan for the next lecture. In the present work for retrieving the posts from the discussion forum we have written the code in PhP and applied keyword extraction techniques on the retrieved results. To compare the extracted keywords, we have written the code using Perl script and identified the percentage match between the human evaluation and keyword extraction technique.

The results laid out in the current study show that the TF-IDF keyword extraction method is the closest technique to human evaluation and can be used instantly to evaluate the post of the discussion forum in a Learning Management System. The present study also presented the various measures of errors and Coefficient of determination of various keyword extraction techniques (see Table 3) between human evaluation and keyword extraction. Furthermore, manual investigation of understanding the learning outcome during online classes for each session is not feasible due to the size and duration of the class. A well-established pedagogical model would help in such investigation and make the teaching and learning effectively in online mode.

6. Declaration

The authors declare that consent from students was collected to use the student data in the current research.

References

[1] Seethamraju R 2014 Effectiveness of Using Online Discussion Forum for Case Study Analysis *Education Research International* 1–10

[2] Vonderwell S 2007. Asynchronous Discussions and Assessment in Online Learning 39(3) 309–328. Retrieved from http://files.eric.ed.gov/fulltext/EJ768879.pdf

[3] Gikandi, J. W., Morrow, D., and Davis, N. E 2011 Online formative assessment in higher education: A review of the literature *Computers and Education* 57(4) 2333–2351. https://doi.org/10.1016/j.compedu.2011.06.004

[4] Moore, R. L., Oliver, K. M., and Wang, C. 2019 Setting the pace: examining cognitive processing in MOOC discussion forums with automatic text analysis *Interactive Learning Environments* 27(5–6) 655–669

[5] Martin, F., Ritzhaupt, A., Kumar, S., and Budhrani, K 2019 Award-winning faculty online teaching practices: Course design, assessment and evaluation, and facilitation *Internet and Higher Education* 42 34–43

[6] Mahrishi M., Morwal S. (2020) "Index Point Detection and Semantic Indexing of Videos—A Comparative Review". In: Pant M., Kumar Sharma T., Arya R., Sahana B., Zolfagharinia H. (eds) Soft Computing: Theories and Applications. Advances in Intelligent Systems and Computing, vol 1154. Springer, Singapore.

[7] Rose, S., Engel, D., Cramer, N., and Cowley, W 2010 Automatic Keyword Extraction from Individual Documents *Text Mining: Applications and Theory* March 1–20. https://doi.org/10.1002/9780470689646.ch1

[8] Gagliardi, I., and Artese, M. T 2020 Semantic unsupervised automatic keyphrases extraction by integrating word embedding with clustering methods *Multimodal Technologies and Interaction* 4(2) 1–20

[9] Aizawa, A 2003 An information-theoretic perspective of tf-idf measures. *Information Processing and Management* 39(1) 45–65

[10] Zhang, W., Yoshida, T., & Tang, X 2011 A comparative study of TF*IDF, LSI, and multi-words for text classification Expert *Systems with Applications* 38(3) 2758

[11] Sparse Distributed Representations (SDRs). (n.d.). Retrieved November 14, 2020 from https://www.cortical.io/science/sparse-distributed-representations/

[12] Govind, Alec C., Spaniol M 2018 Semantic Fingerprinting: A Novel Method for Entity-Level Content Classification. In: Mikkonen T., Klamma R., Hernández J. (eds) Web Engineering. ICWE 2018 *Lecture Notes in Computer Science* 10845 Springer Cham.

Static Vulnerability Analysis of Docker Images

Vipin Jain[1,2], Dr. Baldev Singh[2], Medha Khenwar[1], Milind Sharma[1]
Swami Keshvanand Institute of Technology, Management & Gramothan, (INDIA)
Vivekananda Global University, Jaipur-303012 (INDIA)

ervipin.skit@gmail.com

Abstract: Many organizations are renovating their businesses by grasping DevOps, microservices, and container technologies. Docker is emerged as a new technology, proving an efficient means to develop and deploy applications. Docker containers are created by images to run an application with all its dependencies so that it could run isolated from other processes. Security is always being a foremost concern as our industries are already persistent to improve the reliability and efficiency of new software applications. Security of local Docker containers from the attacks of malicious containers, perceived threats present in Docker images need to be detected and the risks identified when instances of Docker containers run on the host machine. This paper reviews Docker's existing security mechanisms, vulnerabilities, threats and the related tools required for static security analysis.

Keywords: Docker Security; Vulnerability; Container.

1. Introduction

Docker containers are already in demand because it allows the user to run multiple applications on the same host operating system. It also administers the bundle of operating system isolation and security features, but challenge arises when these Docker images are found to be insecure which may cause threats and other major security issues. They should be analyzed statically in order to overcome or render these container images more stable. The above strategy for the container is to be worked before the runtime of the container so that the attack of threats can be recovered without being detrimental to the software [1].

From security aspect, the Docker technology is relatively simple for container images. We need to give priority attention to two layers that can analyze containers in two different ways: Statically or Dynamically. We also usually do not need to concern more about APIs, as the overlay of networks or composite software-defined storage configurations, because these are not a major part of the analyzing portion, as the analysis is done after the container image is built i.e., before the execution of container.

2. Preliminaries

Many scholarly articles are available that are based on the security of Docker containers, but only a few pertaining directly to the static security analysis of Docker containers. Most of the security industries acknowledged that Docker container needs to be the subject of security and has the chances of dissipating. These security concerns are not to be neglected as it may be a big concern to optimize with the emergent of these attacks. This section gives a brief description of the vital terms that are going to be used throughout the paper.

Content from this work may be used under the terms of the Creative Commons Attribution 3.0 licence. Any further distribution of this work must maintain attribution to the author(s) and the title of the work, journal citation and DOI.
Published under licence by IOP Publishing Ltd

IOP Conf. Series: Materials Science and Engineering 1131 (2021) 012018 doi:10.1088/1757-899X/1131/1/012018

2.1 Docker

The "Docker" was in the first instance, founded in March 2013, with the disclosure of a new approach called containerization that led to the virtualization at the OS level. Docker helps to provide a lightweight and quick environment. Docker is sometimes presented as superficial Virtual Machine (VM), but that is not the case. It is totally different from VM. The differences are shown in table 1.

Table 1: Differences between VMs and Docker containers

Virtual Machines	Docker container
It takes a couple of minutes to boot.	Boots in a few seconds.
VMs run using Hypervisor.	Docker makes the use of the Docker engine.
VM tools are straightforward to use and easy to work with.	Docker has a complicated usage mechanism consisting of both third party and Docker managed tools.

Docker gives the aptness to work with the infrastructure in a similar way that applications are handled, adding the image building instructions in "Dockerfile" and sustain a version control of the images. Furthermore, it provides the ability to increase or decrease the resources and scalability to run on different platforms. Containers are not a new abstraction but Docker technology makes its implementation much easier to use. Moreover, while in operation with images it is critical to understand that these images may contain vulnerabilities.

2.2 *Containers*

Docker containers are created by images to run an application with all its dependencies so that it could run isolated from other processes. Containers are hosted in a physical or virtual server on top of their operating system (OS). Containerization is a technology that integrates the application, system libraries, and related dependencies. Eventually, containerization uses host OS which serves relevant libraries and resources. Each container shares the host OS. This makes containers lightweight allowing them to be a few megabytes in size and take a few seconds to start. Containers do not have much management overhead, as they share a common operating system. That system may be patched and fixed if an issue arises [2]. Containers got its name from the shipping industry where different products are placed into compact shipping containers designed to minimize cost and time [3].

2.3 Security

The main concern with the container is security. There are major security issues surrounding containers, such as the security of the containers that are being run on, the content of images by unknown users. It is important to know that the content of container images must be correctly configured. Managing security in containerization is different from other traditional applications. The security challenges occur when an application goes to production. Its chances to be under threat may increase dramatically leading to multiplied user base which implies that it could be open to access from outside the organization [4]. Two types of security checking can be done- static security analysis and dynamic security analysis. Using both the security analysis techniques, it can be checked or confirmed if any of the container images have any bugs. According to the survey conducted in January 2015, 53% of the enterprises revealed that their considerable concern was security. This is because containers rely on the base image and images might contain vulnerabilities which can expand to all containers [5].

2.3.1 Docker Security Scanning

Security scanning has limited features as it can only contrast the software in an image to the familiar vulnerabilities and disclosure database for vulnerabilities [6]. Docker brings forth container images for the limited number of applications/operating systems. These images are of the finest quality for general use but there may be some images that are already in use and are not provided officially by Docker which makes it the major security issue [7]. In the approach of Docker security scanning, there are many scanning techniques already available.

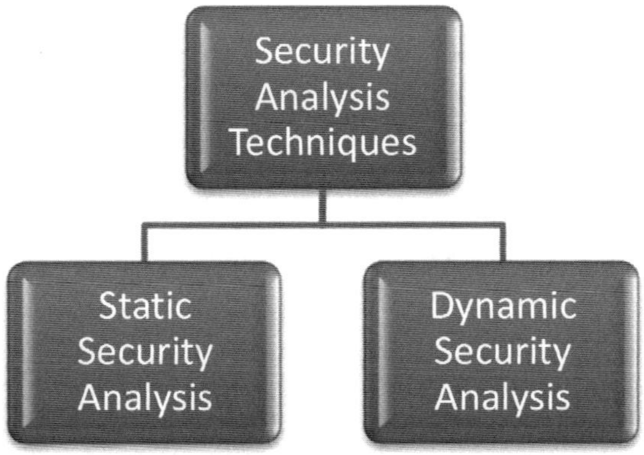

Figure 1: Types of Security Analysis Techniques

2.4 Image Security

Docker has more performance advantages than traditional virtual techniques [8]. There are few areas to consider when reviewing Docker image security like whether its packages are installed in the image or not. When it comes to security, the best choice for a base image is an empty container or distroless (set of images made by Google) or UBI provided by reputed vendors like redhat, IBM, that were created with the intent to be secure [9].

Docker images are- distributed, stored, and managed publically or they can be privately used to build containers. Official images are available on Docker hub used to build public or private processes. Docker images security is an unavoidable concern while creating an infrastructure [10].

- In Docker containers, there is a pre-definition of what is exactly running inside like the path of the data directories, daemon configurations, mount points, etc. A strong focus on security is a must, in this demanding world where the need is not only to make their system fast and reliable but also to make it secure. In 2018, the organizations that use container technology were facing the same issues over security and the percentage of those industries was 60% which is already a huge number that needs to be noticed and cannot be ignored.
- Not only those industries were having the issue of security but with this finding, hundreds of more organizations gave several statistics: They said around 47% of containers are having vulnerabilities and 46% of their developers accept that they had no idea if their containers had vulnerabilities or not.

Some threats and attacks that may cause vulnerabilities in container images and may target the container to fetch their confidential data and make the Docker container less secure are discussed in table 2 below.

Table 2: Docker Container Images Threats with their particular Solution

Threat	Attacks	Explanation	Solution
Cautious Images	Attacks on the containers and host operating system.	Cautious images may cause serious threats and harm the container.	To avoid these attacks, need to use trusted images or registries.
Vulnerable Images	Images may have inbuilt threats.	Some of the images have inbuilt threats that led to the corrupt all the data and damages the whole project.	These vulnerable images are required to follow the analysis process before executing in the container field.
Attack on images	Images that came from unknown sources.	These images may be used as threats and attacks on the container which may target confidential data.	Images need to be verified first either they are secured enough to use or not during build a container.

3. Methodology

In the last few years, various analysis techniques have been used for the security of images with the intention of attaining optimized solutions to get rid of these security issues. In this section, we evaluate these analysis techniques. The methodology adopted in this paper is the result of our work combined with the outcomes from the previous analysis techniques. To our knowledge, the most recent analysis of Docker images was performed by Socchi in 2019. They share the information about security measures introduced by Docker Inc., gives the information of verified and certified images that can improve the security of Docker hub. Besides, they discovered the distribution of all the vulnerabilities across repository types. They implement their software to analyze the image's security. Their conclusion says that the security measures do not improve overall Docker hub security [11].

The container would be able to configure the confidential data of other containers that target the integrity of the application and other information. Furthermore, these containers can also contain similar attacks to another semi-honest container which led another container to be targeted [15]. Our goal is to protect from those threats that may target the other containers and create errors in different ways. An appropriate way to protect those images from different threats and attacks would be to statically analyze the images so that no threats can be detected after the execution process.

For performing the static analysis of a particular number of images, we used an open-source Docker image scanner tool. In this section, we will compare all the analysis that has been already performed during these previous years to find out how the following analysis can be improved in most possible ways. Our findings can be summarized by the table that contains information about all the tools that analyze image security.

According to the surveys that are conducted during previous years, in order to identify all the possible solutions to decrease the vulnerability of images, it must be recognized that the scanning of images is one of the main possible solutions to save our containers from security issues.

4. Discussion

This section discusses the analysis of performance of the software that is already in use to scan Docker hub images or registries. Results are classified as shown in the table with their working environment. There are 10 open source software or tools that are already working to analyze all the trusted or untrusted images to check their vulnerabilities and enhance the security of Docker container images. They are shown in Table 3.

Table 3: Tools to Check Vulnerabilities in Images

Tools	Working	Result
Docker Bench for Security	A Tool to examine Docker container against security specification. It bases its tests on the industry-standard CIS benchmark.	The result conveys the information, pass logs, and warnings.
Clair	Built by CoreOS, performs static analysis of container vulnerability.	By using the Clair API, developers can ask the database for all the issues related to a specific image.
Cilium	Cilium is all about securing network connectivity. A Developer can apply the cilium security policies without making any changes in the application code or container configuration.	Cilium support is great to use, find extensive guides and documentation.
Anchore	A tool for inspecting container security using CVE data and user-defined policies.	It provides a list of vulnerabilities, threat levels, CVE identifiers, and other information.
OpenSCAPWork bench	An environment for developing and maintaining security policies for various platforms.	It allows multiple organizations to efficiently develop security content by avoiding redundancy.
Dagda	A tool for scanning for threats, attacks, vulnerabilities, and malware in Docker containers.	Developers can run it remotely or continuous, it shows the number of threats and other details.
Notary	Notary is a framework enhancing container security with a server of cryptographic responsibility.	It verifies the cryptographic integrity of Docker container images.
Grafes	An API to help and analyze internal security policies.	It helps speed remediation attempts.
Sysdig Falco	It serves behavioral activity monitoring with deep container visibility.	Sysdig provides further container troubleshooting materials.
Banyanops Collector	A framework for static analysis of Docker container images or registries.	It offers deep data analysis.

4.1 Container Image Authenticity

There are many Docker images and repositories on the internet and Docker hub which are doing all kind of useful work, but it comes out with complicated issues while pulling these images without any authenticity mechanism. The questions that relate to the authenticity of images are like:

- Where this image comes from?
- Does this image come from the trusted image creator?
- Does the cryptographic proof say that an author is a person?

- Does the image that has been pulled by you is secured enough to be used?

In any case, Docker allows pulling and running any image or registries by default. Even if the custom images are used during the process, it needs to make sure that nobody inside the organization is able to make changes in an image [4].

Container image authenticity directly implies the security issues of Docker images as the containers can be built by using the same Docker image from the Docker hub or repositories. While discussing the authenticity of Docker images, the first and easiest way to check the vulnerability of images can be done by scanning [12]. The various scanning tools have already been mentioned in table 3. This scanning process scans the vulnerabilities for Docker local images that run on the engine, those local Docker files and images provide users visibility into the security posture. Common vulnerabilities and exposures (CVE) database contain the list of all the vulnerability found during the scanning process [13].

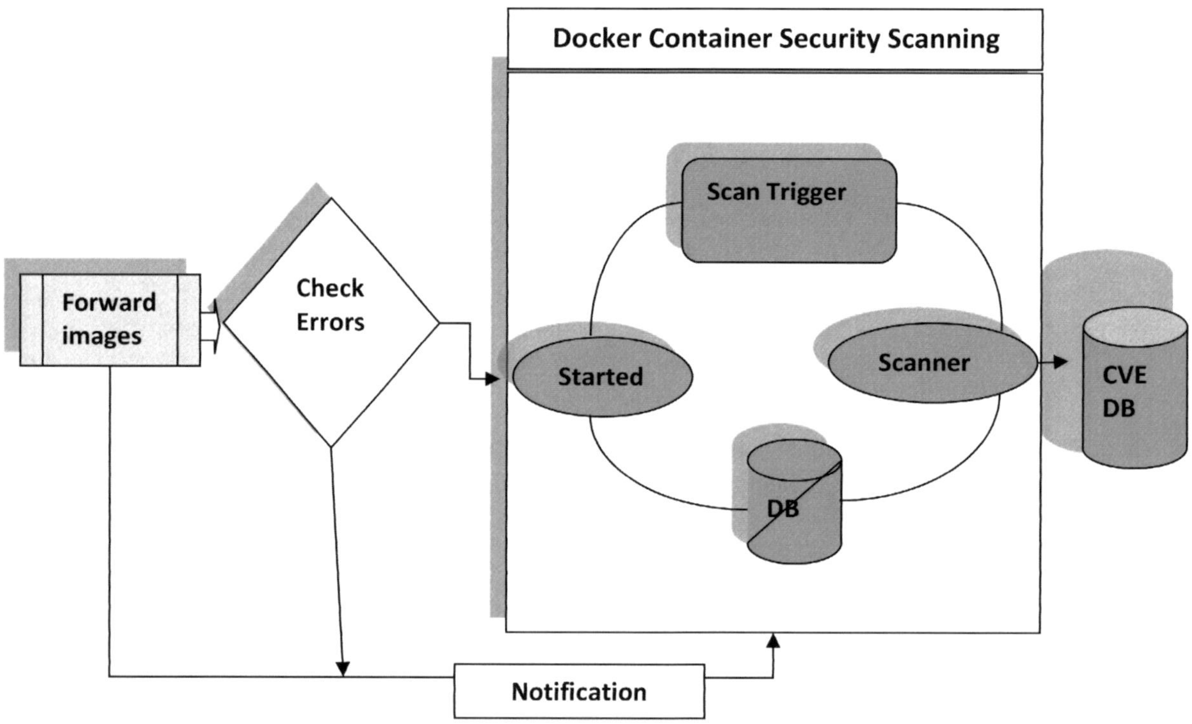

Figure 2: Docker Image Security Scanning

4.2 Vulnerability of Images

In this part, our focus is on the causes that make images more vulnerable as well as how these issues are solved with every other analyzing tool [14]. Below are the root causes behind the vulnerability of Docker images and repositories:

- Insecure generation of images.
- Un-trusted production of image cryptographic configuration.
- Possibility of errors in the image distribution, verification, storage process, and decompression.
- Vulnerabilities inside the images.
- Threats may be directly linked to the Docker or libcontainer.

It is important not to trust any of the images or repositories found on the internet. Docker already sponsors a team dedicated just to review and publish the images only in official repositories. In Docker hub 23% of images are tagged as the latest images and these images are already in the list of most downloaded images from the Docker hub, these images also contain a high amount of vulnerabilities

[16][17][18].

5. Conclusion

In this paper, a review on static security analysis of Docker images is presented. The comparative analysis is being done for security of Docker images using various tools. The security measures are also highlighted that are running with different tools and software created by various platforms to make the containers more secure and reliable. Moreover, creating a secure container images should be chosen prudently so that threats would not become a major concern for containers.

6. References

[1] A. Avram, "Docker: Automated and consistent software deploy-ments," InfoQ. Retrieved, pp. 08–09, 2013.
[2] Abbott, Brendan Michael, "A Security Evaluation Methodology for Container Images" (2017), All Theses and Dissertations 6287, https://scholarsarchive.byu.edu/etd/6287
[3] Shetty, Jyoti. (2017). A State-of-Art Review of Docker Container Security Issues and Solutions. American International Journal of Research in Science, Technology, Engineering & Mathematics.
[4] D. Inc., "What is a container?" https://www.Docker.com/resources/ what-container, access date: 31. Jan 2020
[5] "Docker security," https://docs.Docker.com/engine/security/ security/, access date: 15. Apr 2020.
[6] Y. Zhang, S. Wang, "Research on Docker Security", Network Security Technology & Application, Aug., pp. 32-33, 2017.
[7] Sharma, Mahrishi, Hiran, Doshi, (2020), " Reverse Engineering for potential Malware detection: Android APK Smali to Java" Journal of Information Assurance & Security, Vol. 15 Issue 1, p26-34.
[8] "Cve-2019-9636 detail," https://nvd.nist.gov/vuln/detail/ CVE-2019-9636, 2019, access date: 27. Mar 2020
[9] "Cve-2019-9948 detail," https://nvd.nist.gov/vuln/detail/ CVE-2019-9948, 2019, access date: 27. Mar 2020.
[10] "Cve-2017-1000158 detail," https://nvd.nist.gov/vuln/detail/ CVE-2017-1000158, 2019, access date: 27. Mar 2020.
[11] Adrian Mouat, " Docker Security- Using Docker containers safely", O'Reilly Media, August 2015
[12] E. Socchi and J. Luu, "A deep dive into Dockerhubâ˘ A´Zs security landscape - a story of inheritance?" https://www.duo.uio.no/bitstream/handle/10852/696/ A-Deep-Dive-into-Docker-Hubs-Security-Landscape.pdf? sequence=1&isAllowed=y, 2019, access date: 5. Mar 2019.
[13] D. Huang, H. Cui, S. Wen and C. Huang, "Security Analysis and Threats Detection Techniques on Docker Container," 2019 IEEE 5th International Conference on Computer and Communications (ICCC), Chengdu, China, 2019, pp. 1214-1220, doi: 10.1109/ICCC47050.2019.9064441.
[14] C. Diekmann, J. Naab, A. Korsten and G. Carle, "Agile Network Access Control in the Container Age," in IEEE Transactions on Network and Service Management, vol. 16, no. 1, pp. 41-55, March 2019. doi: 10.1109/TNSM.2018.2889009
[15] S. Kwon and J. Lee, "DIVDS: Docker Image Vulnerability Diagnostic System," in *IEEE Access*, vol. 8, pp. 42666-42673, 2020. doi: 10.1109/ACCESS.2020.2976874
[16] StackRox with observations and analysis from CyberEdge Group "Container and Kubernetes Security: An Evaluation Guide".
[17] Z. Lu, J. Xu, Y. Wu, T. Wang and T. Huang, "An Empirical Case Study on the Temporary File Smell in Docker files," in IEEE Access, vol. 7, pp. 63650-63659, 2019. doi: 10.1109/ACCESS.2019.2905424
[18] Z. Chen, Y. Zhang, and Z. Chen, "A categorization framework for common computer vulnerabilities and exposures," https://academic. oup.com/comjnl/article/53/5/551/415583, 2009, access date: 28. Jan 2020.K. Wist and M. Helsem, "An Extensive Analysis of the Current

Vulnerability Landscape in Docker Hub Images," Master's thesis, Norwegian University of Science and Technology (NTNU), 2020.

An Automated Optimize Utilization of Water and Crop Monitoring in Agriculture Using IoT

Sanwta Ram Dogiwal[1], Pankaj Dadheech[2], Ankit Kumar[3], Linesh Raja[4], Abhishek Kumar[5], Mahender Kumar Beniwal[6]

[1]Associate Professor, Department of Information Technology, Swami Keshvanand Institute of Technology, Management & Gramothan, Jaipur, Rajasthan, India
[2]Associate Professor, Department of Computer Science & Engineering, Swami Keshvanand Institute of Technology, Management & Gramothan, Jaipur, Rajasthan, India
[3]Assistant Professor, Department of Computer Science & Engineering, Swami Keshvanand Institute of Technology, Management & Gramothan, Jaipur, Rajasthan, India
[4]Assistant Professor, Department of Computer Applications, Manipal University Jaipur, Rajasthan, India
[5]Assistant Professor, Department of Computer Science & IT, Jain (Deemed to be University) Bangalore, India
[6]Associate Professor, Department of Computer Science & Engineering, Swami Keshvanand Institute of Technology, Management & Gramothan, Jaipur, Rajasthan, India

[1]dogiwal@gmail.com, [2]pankajdadheech777@gmail.com, [3]iiita.ankit@gmail.com, [4]lineshraja@gmail.com, [5]abhishek.maacindia@gmail.com, [6]mbeniwal@gmail.com

Abstract. AI Solution for Farmers is an agricultural based project, created to help farmers and to help them in increasing their productivity. Since the technologies are running the world, why agriculture must be deprived of it. Agriculture is one of the most important areas that have major impacts on the economy and society of a country. Technological developments serve as tools to share knowledge and practices of agricultural products and make more satisfactory lives for farmers, traders, policymakers, and the overall society. It is evident that knowledge has become a crucial component in production, society, food security, health, poverty, and other millennium development goals. The use of technology may provide a better approach to solve the problems arising from sowing the seed to harvest the crop. The new technologies like Machine Learning and Data Science may be great help to have an eye on the deciding factors to the growth of crop.

Keywords: IoT, Sensors, Raspberry-PI, Microcontroller, Wi-Fi.

1. Introduction

An intelligent system to utilize internet of things (IoT) to monitor a plurality of crops in an agriculture field and further enables a farmer to manage the crops and the agriculture field in real-time is provided substantially, as shown in and/or described in connection with at least one of the figures, as set forth more completely in the claims.

Content from this work may be used under the terms of the Creative Commons Attribution 3.0 licence. Any further distribution of this work must maintain attribution to the author(s) and the title of the work, journal citation and DOI.
Published under licence by IOP Publishing Ltd

The intelligent system comprises a plurality of sensor units, a wireless unit, a server unit, a database, a computing device, a software application, a water motor and water management unit, a buzzer unit, a CCTV camera, an unmanned-aircraft vehicle system (UAVs), and a pesticides control unit and crop quality unit [1]. The plurality of sensor units are configured with the crops to periodically detect a plurality of parameters pertaining to the crops and the agriculture field. The plurality of parameters include but not limited to temperature and humidity around the agriculture field, an obstacle in the agriculture field, soil condition and a plurality of predefined vibrations frequencies.

In an aspect, the plurality of sensor units comprises a DHT temperature and humidity sensor, an ultrasonic sensor, a soil moisture and humidity sensor, and a fence sensor [2]. The DHT temperature and humidity sensor detects the temperature and humidity around the agriculture field. The ultrasonic sensor detects the obstacle in the agriculture field to provide security to the agriculture field and further detects the water level in a tank placed in the agriculture field.

The soil moisture and humidity sensor detects the soil condition. The soil moisture and humidity sensor is configured with a water motor and water management unit to initiate a command to supply water. The fence sensor detects a plurality of predefined vibrations frequencies generated by the movement of an animal or a human, cutting, climbing or lifting around a fence of the agricultural field to protect the agricultural field and further transmits a notification signal to the computing device of the farmer [3].

The wireless unit is configured with the sensor units to transmit the detected data to a remotely placed server unit via a communication network. The server unit is configured with a database to store the detected data. The computing device connected to the server unit to receive the detected data from the server unit [4]. The software application is integrated with the computing device to provide a plurality of indicators pertaining to crops and weather forecast to manage the agriculture.

The buzzer unit is configured with the fence sensor to generate an audio signal on detection of the vibrations to alert the farmer. The CCTV camera and an unmanned-aircraft vehicle system (UAVs) for continuously monitoring the agriculture field and transmitting the monitored to the computing device of the farmer via the server unit. The pesticides control unit and crop quality unit provides a predefined ratio of pesticides to prevent the crops from insects. The software application [5] provides the detected data in an analog form and enables the farmer to check the quality of their corps.

In an aspect, the water motor and water management unit automatically supplies water to the soil on receiving a wireless command from the computing device of the farmer. The ultrasonic sensor monitors and controls the water level in a water for pouring the water onto the soil when required [6].

2. Literature Review

This monitoring systems can be classified according to the environment that is used for, such as industrial, home, office, agriculture and others environment. In literature system, this smart agriculture monitoring system project will be discussed among the previous research of a project [7]. Selected publications will be a guide for reference in this literature review. The table below shows the developed system's critical review. These projects have successfully come on board because they target a specific location only. This smart agriculture monitoring system is inexpensive and efficient project. It is better than the hazardous traditional methods.

Table 1. Critical Survey on Development System

S. No.	Functionality	[1]	[2]	[3]	[4]	[5]
1.	User Friendly	Yes	Yes	Yes	Yes	Yes
2.	Low Cost	Yes	Yes	No	No	No
3.	Application [Mobility]	Yes	Yes	WSN	WSN	Sensor
4.	Power	Yes	LOW	Low	Low	Low

5.	Employ	Home Control	Agriculture Industry	Agriculture Industry	Agriculture Industry	Agriculture Industry
6.	User Interface	Yes	-	Yes	Yes	Yes
7.	Security	Yes	-	Yes	Yes	-
8.	Load Limit [web server]	Yes	Yes	Yes	Yes	No
9.	Simplicity	Yes	-	No	No	Yes
10.	Flexibility	Yes	Yes	Yes	Yes	Yes
11.	Community	Yes	Yes	Yes	Yes	No
12.	Freedom	Yes	Yes	Yes	Yes	Yes
13	Easy installation & Upgrade	Yes	Yes	No	Yes	Yes
14	Display control	Yes	-	Yes	Yes	Yes
15	Configure Website	Yes	Yes	Yes	Yes	No

3. Proposed Methodology

Recent time farmers have used irrigation technique in which the farmer manually switch the on and off the water-pump according to the need. Besides indicating power points, the ballasts do not indicate on the load, thereby farmers cannot know the power point is low [8, 9]. If they switch OFF any motor, the sudden slowdown of armature circuit. They may have to travel too far to plug and unplug the unit from electricity. They may be experiencing the heat, rain and at night too. Upon reaching the farm, they found out that there is no power and they sighed in disappointment.

3.1 Working of Proposed Model

- Our model is responsible for obtaining data regarding the location, time and weather conditions (most prominently rain) and automate the working of irrigation system according to the moisture content of the soil at required intervals in a day.
- The model facilitates agility while working with varied soil types, crop variants in different weather conditions and seasons [10].
- All of this is done while keeping in mind its feasibility to farmers even in the remote areas.
- Our smart controller enables users to control and monitor their home's irrigation through mobile smartphone. Thee data collected by sensors is sent to the database created on the cloud for further processing and future weather condition prediction purposes.

In the app farmer have to login to register with the authorized login [11]. The authentication location has to submit with their relative field of strength. The login has to commit with the authorized token. The authentication token is sent to the own relative id to the relative society to the end of submission.

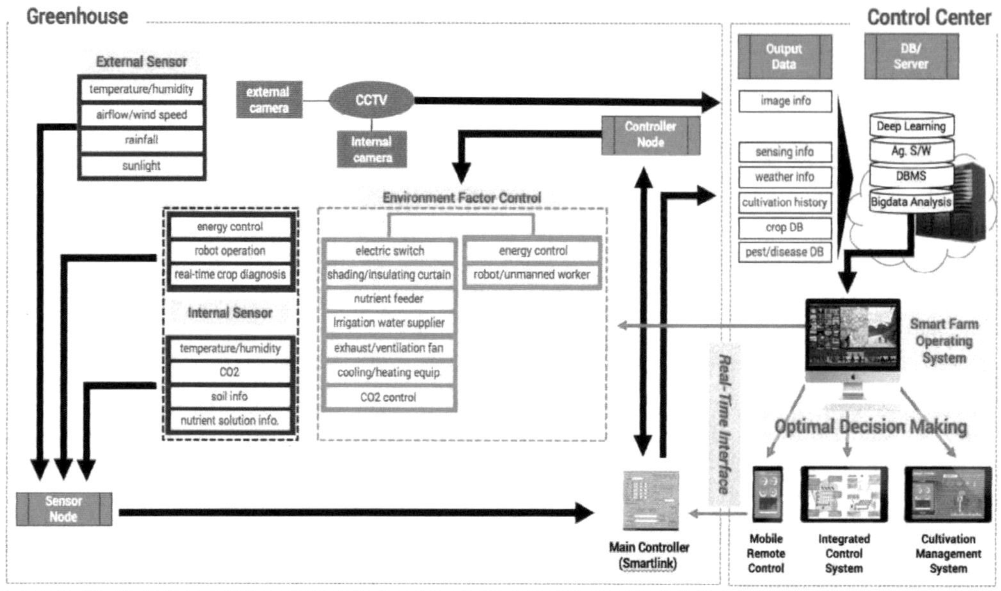

Figure 1. ERD diagram of sprint backlog

Smart farm operating system is a software consists of mainly three components mobile remote control, integrated control system and cultivation management system. This smart system is based on machine learning, by the help of which one can find suitable crop, environment for cultivation with the help of some information based on images, weather and cultivation history stored in the database. Basically System's main controller interacts with the three major nodes: sensor node, controller node and the mobile remote controller [12] .Here external sensor node senses all the external environmental activities such as temperature, humidity, wind speed, rainfall and sunlight. Another major sensor is internal sensor node which detects the inside information of soil such as humidity, temperature, nutrient information, CO2 content of soil, this whole data coming from external and internal sensor node then sent to the main sensor node which makes the right information out towards the main controller. Main controller then instructs the environment controller (EC node) node based on the data it is gaining from the sensor node, EC node automatically turn electric switches (motor, electricity, etc.) on/off, EC node also perform other activities which totally replaces the manual interaction such as shading/insulating curtain, irrigation water supply, exhaust/ventilation fan [13], temperature control (heating/cooling), CO2 control and these all activities are monitored in CCTV camera.

4. Experiments & Results

An experiment for the smart irrigation system was made on an ornamental plant grown in a greenhouse. Plant requires 1200mm of water per day and its soil temperature should be not less than 50°C. In the Arduino program, using variables moisture and temperature, they were set to 300 and 450 respectively. This method is an effective way of saving water, inexpensive and efficient.

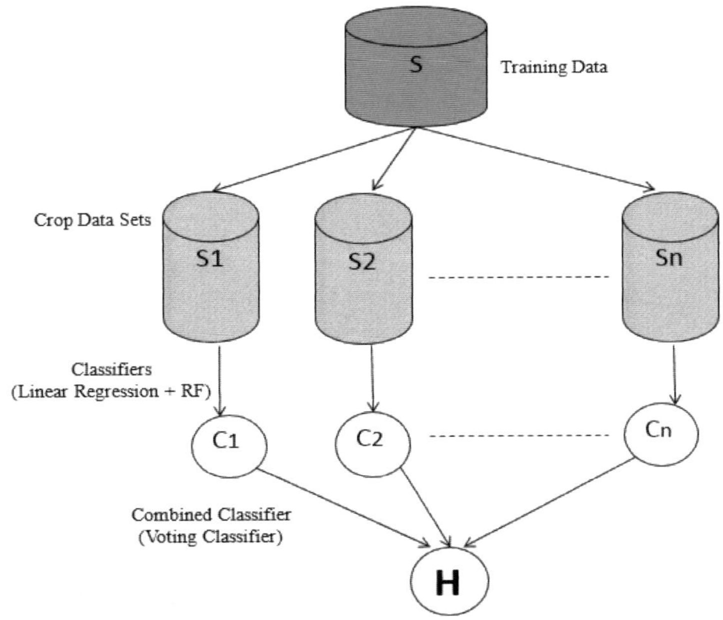

Figure 2. Proposed Model Diagram

In Smart Agriculture system it shows the real time data by sensing the agriculture field data using Wi-Fi as allows user to control the system and monitor their agriculture from anywhere of world as the application is connected to the cloud where the data from sensor is stored and allow application to connect to hardware. There uses various sensors like ,DHT (Temperature and humidity) sensor [14] for sensing the local atmosphere data, Ultrasonic sensor for sensing the distance or obstacle in the field as it provide security to the agriculture field as well as for water level indicator in the tank, Soil moisture and humidity sensor for sensing the current soil condition, water motor for supplying the water to the soil automatically by sensing through sensor, Fencing of field for security to protect the field from animals or theft as it will send notification to the farmers mobile and buzzer will start buzzing ,CCTV camera and Drone technology for continuously monitoring of farm through application, Pesticides control system [15] which provide the perfect solution from insects for their valuable crops. Proper ratio of pesticides is given to the farmer so that they can take action quickly and accurately. The application is used to control and monitor data sensed from sensor and send it to server then to user application which is connected to hardware through internet system. The android application keep the data for weeks and even for months.

We know that Indian economy is mostly depends on agriculture but the farmer who produce crops are living a poverty life and are not able to feed their family as they are considered as the god but they itself are living a beggar life. Due to which sometimes they do suicide by getting frustrated from their life. This is due to the fact that they are not able to grow quality crops and their values in the market is low because they do not provide required material to crop at the time due to which it get destroyed because of not able to get correct prediction [16]. The prediction include such as watering or pouring water to plant or soil which get wrong mostly and because of this the problem arises.

4.1. Performance Evaluation

Every instance is analysed in two categories in the binary classification model [17]. There are two categories, namely, true and false. True positive is the number of correct predictions that an instance is positive. False positive (FP): The number of incorrectly predicting instances as positive. False Positive (FP): The number of correct predictions that an instance is positive. The number of correct predictions that this sentence is negative.

IOP Conf. Series: Materials Science and Engineering 1131 (2021) 012019 doi:10.1088/1757-899X/1131/1/012019

Table 2. Representation of Confusion Matrix also called Contingency Table

	Observation	
Experiment	True	False
True	TP	FP
False	TN	FN

An uncertainty matrix shows the associations between observable and expected in a certain phenomenon [18]. Along the major diagonal of the table, the correct classification should be shown as true positives and true negatives. The model errors are denoted by the additional columns. A perfect math model would contain only the true positive and true negative values. From the confusion matrix, various indicators can be derived. Accuracy is the overall success rate of the classifier.

$$Accuracy = (TP + TN) / (TP + FP + FN + TN) \quad (1)$$

False Accept Rate (FAR):
It is the ratio of no of accepted data in the dataset by the total data present in the dataset

$$FAR = \frac{No\ of\ DataAccepted\ out\ side\ the\ dataset}{Total\ No\ of\ Data\ in\ Dataset} \quad (2)$$

False Reject Rate (FRR):
It is the ratio of false selected data for verification by the total data present in the system

$$FRR = \frac{No\ of\ Correcte\ Data\ Reheceted}{Total\ No\ of\ Data\ in\ Dataset} \quad (3)$$

True Success Rate (TSR): This is the relationship between the exact number of the database and the total number of total population in the database.

$$FRR = \frac{No\ of\ Correcte\ Matched\ Data}{Total\ No\ of\ Data\ in\ Dataset} \quad (4)$$

Table 3. Production: Year v/s Area

S. No	Crop_Year	Area	Production
0	1997	45500.0	156800.0
1	1998	48730.0	176455.0
2	1999	52097.0	139835.0
3	2000	31501.0	115721.0
4	2001	44550.0	176784.0

Table shows that annual production of all crops in Rajasthan State year wise. We have separated the Production of crop district wise.

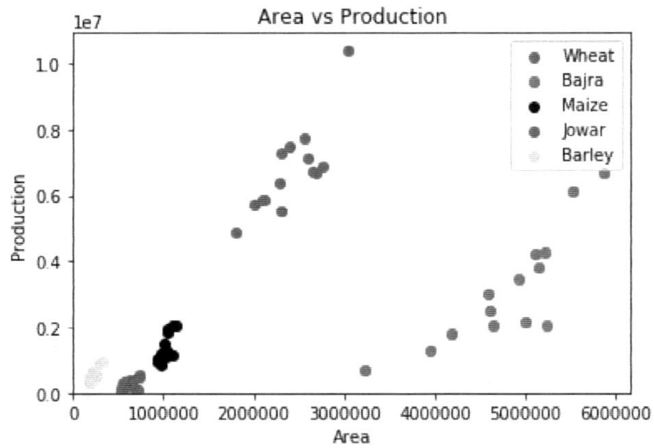

...

Figure 3. Production of Crop v/s Area

Description: The above chart displays the observation points between Production Area and Total Production for various crops like Wheat, Bajra, Maize, Rice and Barley from the previous year's Statistical dataset [19, 20, 21]. Different colors are used to show the observation points for different crops. In the Statistical Analysis of existing data we can see that Wheat and Bajra needs larger production area as compared to other crops.

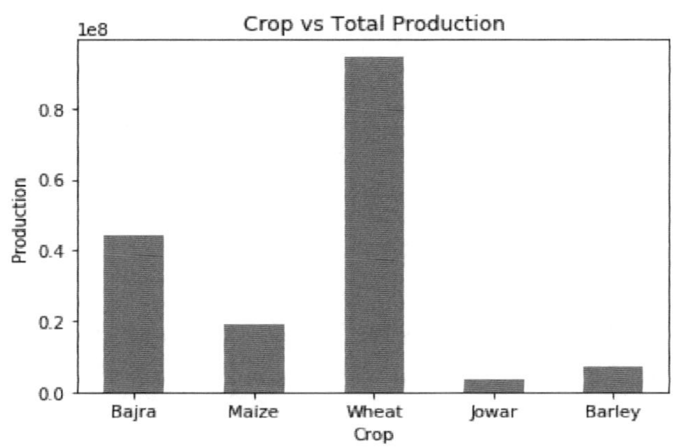

Figure 4. Production of crop

Description: The above chart displays the relation between crop and their production for various crops like Wheat, Bajra, Maize and Barley etc. from the previous year's dataset [22]. The Statistical Analysis of existing data in above chart shows that the production of wheat, Bajra and Maize is highest than other crops that's why we can say that Wheat, Bajra and Maize crops have the more suitable environment in Rajasthan [23].

Description: Different models are trained with the common dataset and then these models are used for crop yield prediction and their accuracy of prediction is calculated as per below details:

a) Linear Regression – Linear model is created using Python library and trained it with real time agriculture dataset. Accuracy of the model is calculated 63.73%
b) Ridge Regression - Ridge Regression model is created using Python library and trained it with same agriculture dataset. Accuracy of the model is calculated 63.73%
c) Random Forest Regression – This model is based on the concept of decision trees. It creates the multiple decision tree for predicting the output and finally combine the result of all decision trees to get the final prediction. This model is also trained with the same agriculture dataset and Accuracy of the model is calculated 87.08

d) Proposed Hybrid Approach – Above mentioned model are good for machine learning purpose but may not give the high accuracy for huge dataset. So in order to achieve the high accuracy we used the hybrid model approach.

This approach merges the concept of linear, Ridge and Random Forest model and creates a hybrid model with the improved accuracy calculated 93.72. In the proposed work accuracy [24, 25] is very important factor, model with the high accuracy can precisely predict the production of specific crop in the specific location and the same can be informed to farmer that what amount of crop can be produced in his/her production area. Balanced production of the crops will not affect the production of other suitable crops in the same location and this way it will not lead the increment in the cost of crop due to shortage of production and will fulfil the need of every people within the cost.

5. Conclusion and Future Scope

With fully equipped software and Internet of Things, agriculture industry can provide a better vision for next generation and make India better in coming days. It is an application based on machine learning and IOT technology. It is designed to help farmers to find most suitable crop according to their soil, climate condition of area, rainfall and production in the market to attain more profit and use minimum fertilizers. Therefore, the crop quality will definitely improve. This web application is very interactive as it uses face detection for login and use speech API's for interaction. Therefore, it is very easy to use, much interactive, and easily understandable by farmers. The basic and only need is the internet connection to use the application. There is a tremendous future of IoT in the field of agriculture, whether it is in innovation or finance. If farmers are provided with proper training about technologies, they will be able to perform much of their agricultural tasks through mobile. It allows farmers to stay in contact with their farms from anyplace. It simultaneously reduces human effort with increased productivity and boosts the economy of farmers. This plan will completely revolutionize the agriculture industry in India.

6. References

[1] Ngu, A.H., Gutierrez, M., Metsis, V., Nepal, S., Sheng, Q.Z., IoT middleware: a survey on issues and enabling technologies. *IEEE Int. Things J.* 4(1), 1–20 (2017)

[2] Khattab, A., Abdelgawad, A., Yelmarthi, K., Design and implementation of a cloud-based IoT scheme for precision agriculture. In: *28th International Conference on Microelectronics*, pp. 201–204 (2016)

[3] Bing, F.: The research of IOT of agriculture based on three layers architecture. In: *2nd International Conference on Cloud Computing and Internet of Things*, pp. 162–165 (2016)

[4] Mahrishi, M., Hiran, K. K., Meena, G., & Sharma, P. (2020). Machine Learning and Deep Learning in Real-Time Applications, IGI Global.

[5] Edwards-Murphy, F., Magno, M., Whelan, P.M., O'Halloran, J., Popovici, E.M., B+WSN: smart beehive with preliminary decision tree analysis for agriculture and honey bee health monitoring. *Comput. Electron. Agric.* 124, 211–219 (2016). https://doi.org/10.1016/j.compag.2016.04.008

[6] Vernandhes, W., Salahuddin, N.S., Kowanda, A., Sari, S.P., Smart aquaponic with monitoring and control system based on IoT. In: *2017 Second International Conference on Informatics and Computing (ICIC)*, pp. 1–6. IEEE (2017)

[7] Yang, K., Liu, H., Wang, P., Meng, Z., Chen, J., Convolutional neural network-based automatic image recognition for agricultural machinery. *Int. J. Agric. Biol. Eng.* 11(4), 200–206 (2018)

[8] Johannes, A., et al., Automatic plant disease diagnosis using mobile capture devices, applied on a wheat use case. *Comput. Electron. Agric.* 138, 200–209 (2017). ISSN: 0168-1699. https://doi.org/10.1016/j.compag.2017.04.013.

[9] Mr. Ankit Kumar, Dr. Dinesh Goyal, Mr. Pankaj Dadheech, (2018), "A Novel Framework for Performance Optimization of Routing Protocol in VANET Network", *Journal of Advanced*

Research in Dynamical & Control Systems, Vol. 10, 02-Special Issue, 2018, pp-2110-2121, ISSN: 1943-023X.

[10] Mr. Pankaj Dadheech, Dr. Dinesh Goyal, Dr. Sumit Srivastava, Mr. Ankit Kumar, (2018), "A Scalable Data Processing Using Hadoop & MapReduce for Big Data", *Journal of Advanced Research in Dynamical & Control Systems*, Vol. 10, 02-Special Issue, 2018, pp-2099-2109, ISSN: 1943-023X.

[11] Dhaka, V. S., Poonia, R. C., & Raja, L. (2014),"The Realistic Mobility Evaluation of Vehicular Ad-Hoc Network for Indian Automotive Networks", *International Journal of Ad hoc, Sensor & Ubiquitous Computing*, 5(2), 1.

[12] Poonia, R. C., & Raja, L. (2018), "On-Demand Routing Protocols for Vehicular Cloud Computing", *Vehicular Cloud Computing for Traffic Management and Systems*, (pp. 151-177). IGI Global.

[13] Pankaj Dadheech, Dinesh Goyal, Sumit Srivastava & C. M. Choudhary, (2018), "An Efficient Approach for Big Data Processing Using Spatial Boolean Queries", *Journal of Statistics and Management Systems (JSMS)*, 21:4, 583-591.

[14] A. Kumar and M. Sinha (2014), "Overview on vehicular ad hoc network and its security issues," *International Conference on Computing for Sustainable Global Development (INDIACom)*, pp. 792-797. doi: 10.1109/IndiaCom.2014.6828071.

[15] Pankaj Dadheech, Ankit Kumar, Chothmal Choudhary, Mahender Kumar Beniwal, Sanwta Ram Dogiwal & Basant Agarwal (2019), "An Enhanced 4-Way Technique Using Cookies for Robust Authentication Process in Wireless Network", *Journal of Statistics and Management Systems*, 22:4, 773-782, DOI: 10.1080/09720510.2019.1609557.

[16] Ankit Kumar, Pankaj Dadheech, Vijander Singh, Linesh Raja & Ramesh C. Poonia (2019), "An Enhanced Quantum Key Distribution Protocol for Security Authentication", *Journal of Discrete Mathematical Sciences and Cryptography*, 22:4, 499-507, DOI: 10.1080/09720529.2019.1637154.

[17] Ankit Kumar, Pankaj Dadheech, Vijander Singh, Ramesh C. Poonia & Linesh Raja (2019), "An Improved Quantum Key Distribution Protocol for Verification", *Journal of Discrete Mathematical Sciences and Cryptography*, 22:4, 491-498, DOI: 10.1080/09720529.2019.1637153.

[18] Ankit Kumar and Madhavi Sinha (2019), "Design and analysis of an improved AODV protocol for black hole and flooding attack in vehicular ad-hoc network (VANET)", *Journal of Discrete Mathematical Sciences and Cryptography*, 22:4, 453-463, DOI: 10.1080/09720529.2019.1637151.

[19] Ankit Kumar, Vijayakumar Varadarajan, Abhishek Kumar, Pankaj Dadheech, Surendra Singh Choudhary, V.D. Ambeth Kumar, B.K. Panigrahi, Kalyana C. Veluvolu, "Black hole attack detection in vehicular ad-hoc network using secure AODV routing algorithm", *Microprocessors and Microsystems*, 2020, 103352, ISSN 0141-9331, https://doi.org/10.1016/j.micpro.2020.103352.

[20] Ankit Kumar, Linesh Raja, Pankaj Dadheech, Manish Bhardwaj (2020), "A Hybrid Cluster Technique for Improving the Efficiency of Colour Image Segmentation", *World Review of Entrepreneurship, Management and Sustainable Development*, Nov. 2020, Vol. 16, Issue 6, pp. 665-679, Print ISSN: 1746-0573 Online ISSN: 1746-0581, https://doi.org/10.1504/WREMSD.2020.111405.

[21] Dadheech P., Goyal D., Srivastava S., Kumar A., Bhardwaj M. (2021) Performance Improvement of Heterogeneous Cluster of Big Data Using Query Optimization and MapReduce. *In: Goyal D.,*

Bălaş V.E., Mukherjee A., Hugo C. de Albuquerque V., Gupta A.K. (eds) Information Management and Machine Intelligence. ICIMMI 2019. Algorithms for Intelligent Systems. Springer, Singapore, pp 85-100, First Online: 17 September 2020, Print ISBN: 978-981-15-4935-9, Online ISBN: 978-981-15-4936-6. https://doi.org/10.1007/978-981-15-4936-6_9.

[22] A. Kumar, P. Dadheech and U. Chaudhary, (2020), "Energy Conservation in WSN: A Review of Current Techniques," *2020 3rd International Conference on Emerging Technologies in Computer Engineering: Machine Learning and Internet of Things (ICETCE), Jaipur, India*, 2020, pp. 1-8, ISSN: 978-1-7281-1683-9, doi: 10.1109/ICETCE48199.2020.9091736.

[23] Mahrishi M., Morwal S. (2020) "Index Point Detection and Semantic Indexing of Videos—A Comparative Review". In: Pant M., Kumar Sharma T., Arya R., Sahana B., Zolfagharinia H. (eds) Soft Computing: Theories and Applications. Advances in Intelligent Systems and Computing, vol 1154. Springer, Singapore.

[24] Pankaj Dadheech, Ankit Kumar, Vijander Singh, Linesh Raja, Ramesh Poonia (2020), A Neural Network-Based Approach for Pest Detection and Control in Modern Agriculture Using Internet of Things. *In: Amit Kumar Gupta, Dinesh Goyal, Vijander Singh, Harish Sharma (eds), Smart Agricultural Services Using Deep Learning, Big Data, and IoT*, October, 2020, pp 1-19, ISBN13: 9781799850038, ISBN10: 179985003X, EISBN13: 9781799850045, DOI: 10.4018/978-1-7998-5003-8.ch001, Publisher IGI Global.

[25] Ankit Kumar, Pankaj Dadheech, Vijander Singh, Linesh Raja (2020), "Performance Modeling for Secure Migration Processes of Legacy Systems to the Cloud Computing", *In: Tin Thein Thwel, G. R. Sinha (eds), "Data Deduplication Approaches: Concepts, Strategies and Challenges", Chapter-13*, pp. 255~280, ISBN: 978-0-12-823395-5, DOI: https://doi.org/10.1016/B978-0-12-823395-5.00003-3, Publisher Elsevier.

Multiple Watermarking Scheme for Video & Image for Authentication & Copyright Protection

Neha Janu[1], Ankit Kumar[2], Pankaj Dadheech[3], Gajanand Sharma[4], Ashutosh Kumar[5], Linesh Raja[6]

[1]Associate Professor, Department of Computer Science & Engineering, Swami Keshvanand Institute of Technology, Management & Gramothan, Jaipur, Rajasthan, India
[2]Assistant Professor, Department of Computer Science & Engineering, Swami Keshvanand Institute of Technology, Management & Gramothan, Jaipur, Rajasthan, India
[3]Associate Professor, Department of Computer Science & Engineering, Swami Keshvanand Institute of Technology, Management & Gramothan, Jaipur, Rajasthan, India
[4]Assistant Professor, Department of Computer Science & Engineering, JECRC University, Jaipur, Rajasthan, India
[5]Assistant Professor, Department of Computer Science & Engineering, JECRC University, Jaipur, Rajasthan, India
[6]Assistant Professor, Department of Computer Applications, Manipal University Jaipur, Rajasthan, India
[1]neha123jaipur@gmail.com, [2]iiita.ankit@gmail.com, [3]pankajdadheech777@gmail.com, [4]gajanan.sharma@gmail.com, [5]ashucse007@gmail.com, [6]lineshraja@gmail.com

Abstract. Watermarking has been used frequently to authenticate the accuracy and security of the image and video files. In the world of computer technology, several watermarking strategies have developed during the past 20 years. Integrate a picture of identity that is not always concealed, such that no detail is not possible to delete. A monitoring code can also be used to deter unauthorized recording equipment. Another application is the watermark copyright control, which works at stopping the creator of the image from stealing photos unlawfully. A watermark is a promising option for the copyright transcendence of multi-media files, as embedded messages are still included. Due to the limits of fidelity, a watermark may be implemented in a small multimedia data space. There is no proof that Automated Watermarking technologies will fulfil the ultimate purpose of the cleaners of all sorts of copyright security operations to gather knowledge from the data they obtain. Relevant situations may be deemed more fairly expected with the usage of automated copyright marking technologies. A perfect device will not be able to add a digital watermark without the limit, which does not supply the whole object with details. In this work, a modern technique for watermarking includes injecting two or more messages or photographs into a single picture for protection purposes and repeating a similar procedure for N-frames for authentication in the film.

Keywords: Digital Watermarking, Video, Authentication, Temper, Copyright, Protection, DCT.

Content from this work may be used under the terms of the Creative Commons Attribution 3.0 licence. Any further distribution of this work must maintain attribution to the author(s) and the title of the work, journal citation and DOI.
Published under licence by IOP Publishing Ltd

1. Introduction

There are many interactive items in one. The basic definition of multimedia is "the fusion of two or more media." The media in multimedia are displayed in several ways: graphics, image, text, sound effect, audio, voicing, etc. Both symbolic and functional purposes, each serves as an effective medium for communication. The media will have a more diverse and interactive experience when melded together. In the event of collaboration and communication between the different media elements the subsequent steps can be strengthened even further. The famous sentence, "The medium is the message", has been used by Marshall Mc Luhan as a leading and prominent media relations theorist. According to him the attention should not be on the information or what is being said, but the mechanism by which it is provided. He felt it was the way that forms and governs the size and form of human association and behaviour. The topic is not meaningless, but the format of transmission is significant in the way the message is conveyed. This is where multimedia has the incredible power and impact. Media is the plural medium by choice. The word 'facilitates or links contact' has evolved, whether by a phone, the internet, the TV or another instrument. It is instant and does not require consultation to talk openly to those on one side. This is the purest mode of contact. A medium's function is to enable the message to be conveyed. We call multimedia, whether or not it's computer-based, when we use more than one form of media. At the moment, the media used news and information mostly in the press to transmit to the masses. Now there are a number of ways of contact in the media.

Multimedia is a synergistic process in which many elements of the media operate together to strengthen and unify a whole. A combination of media adds wealth and offers a total sensory experience. Multimedia was once the simultaneous replay of a slide and a tape storage unit. As an example, 50 years ago the audio tried to synchronize with the series or play as 'bottom' music, photographic photographs in slide form were projected on a computer.

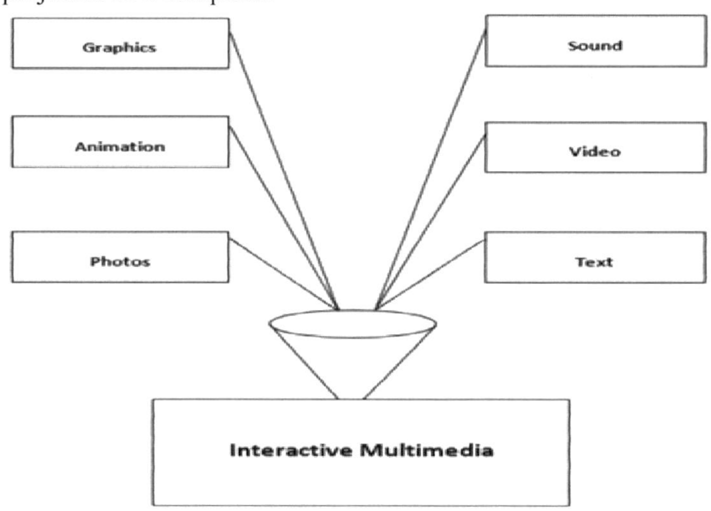

Figure 1. Disparate Multimedia Elements Funnelling Into One Unit

Multimedia uses various media formats (such as text, sound, graphics, animation, film, interactive) in the collection and processing of information to educate or entertain the user of these. The use of electronic devices to store and experience visual content also applies to multimedia. Multimedia is comparable, but larger, to conventional mixed media art. "Rich media" means immersive multimedia.

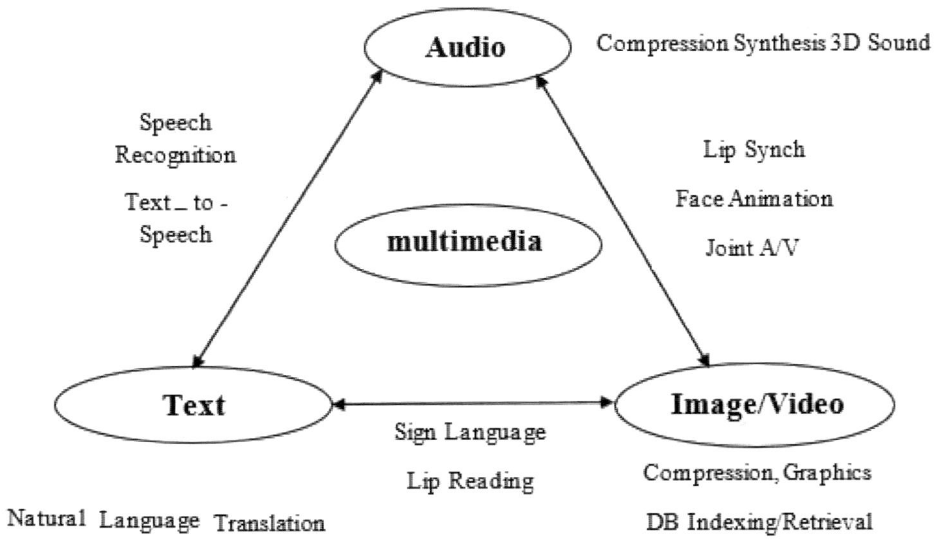

Figure 2. Multimedia uses various media formats

2. Literature Review

A New H.264 / AVC watermarking method is proposed [1]. They were embedded robust and fragile watermark into DCT domain to the motion vector, which can work together to achieve copyright protection and authentication of both sides. Our solution is superior to higher capacity and another video watermark plan, especially in low bit-rate compression. In addition, being a good mode selection function in H.264 / AVC Lagrangian optimization aligned. The software would only trigger very minor changes to the material. The solution is very efficient algorithm to extract watermark from emails.

Such act ensures the creators of copyrighted works have actual power over the contents of their archives. We wish to apply DW-SVD algorithm in creating digital watermarks [2]. The proposed method includes two steps: the embedding of watermark and the extraction of watermark. In the process of embedding and embedding, we mix DCT and CSS embedding methods with PSNR and Video Concealment. "Watermarking method based on DWT and SVD proposed hybrid concept of digital video". This technique can be used to hide information from people. DCT is rather time-consuming method because it requires another database. A hybrid result was found superior over a single DWTSVD method. It was found that the watermarking method met the required criteria. It is of fast processing and is probably robust.

Hiding video data is a big research topic [3]. They suggest a new video data hiding method which uses repeated error correction to accumulate area code penalty and supremacy of data hiding (FZDH). FZDH is authorised for no modification and data hiding method. The frame was checked for any form of file, including.MP4,.3gp,.AVI etc., and the performance of the whole method of hiding video output details. Protection of the proposed scheme, video encryption and decryption secret methods. The simulation results indicate that the surveillance video phase is secret.

Video watermarking is a recent technology, primarily used for the illicit spread of multimedia video on the internet[4]. When a watermark is added to the recording, it must not be detected when sound is still visible. The encoded original picture details can be transparent or invisible as a „digital watermark." More nuanced attacks can be carried out to enhance the protection and method of embedding watermarks. Research has shown that the most effective and economic way to evaluate Eigenwerte is by utilising the SVD algorithm. For purposes of embedding and extraction method, the advanced SVD transformation algorithm suggested in this work would be. The experimental findings suggest that the watermarking method suggested for SVD algorithms is comparatively less risky.

The plans are focused on the copyright security of validity and discrete wavelet transformation (DWT) and single-value decomposition (SVD). These are grey watermark logo image. As the watermark, the algorithm that the original image was converted into the wavelet domain and the use of a reference sub-picture, in which a contrast is set and value coefficients of a wavelet are used and shaping. In this way, LUCC can have a significant influence on political and economic decisions.

Proposed work scans the image for the digital image obtained from the watermarked software. In their algorithm, cover pictures they use as a watermark [6]. It combines object watermarking techniques with blurry block images. It also embeds the watermark in the host image displacement and LSB. A watermark image is copied from the suspect image and inverted.

The use of PKI, public key encryption and watermarking technologies in developing a visual image of the latest testing and evaluation methods that apply to information system security [7]. The main idea of their paper is that the image was watermarked (LSB).

Suggested to apply zero-knowledge cryptography for secure the watermark picture [8]. In its scheme using DWT (discrete wavelet transforms) of the embedded watermark, they made adequate secure watermark in the bit least significant in host image (LSB).

Proposed a fragile watermark image authentication scheme [9]. They used the singular values of the image decomposition (SVD) singular values, integrity check images. In order to make the singular values authentication data is changed to use modular arithmetic bit. Then, they are inserted least significant bits of the original image bit (LSB). The change in the original image pixels are selected randomly.

Proposed a new standard of structural digital watermarking by using the random function mapping for LSB [10]. They believed that the randomized position of the image is more reliable that the use of traditional hashing algorithms.

Proposed compromise between concealment and using SSIM quality measure of watermark robustness [11]. They detect watermarked images using complex statistical procedures.

This proposed a bit-plane index modulation based on fragile watermarking scheme used to verify the RGB colour image [12]. By embedding the R, G, in the R, G, B component of the watermark image, the original image of the B component, and as the embedding distortion is minimized by using the least significant bit (LSB) to change the program. Their encryption method includes encoding and decoding procedures. We hope to deal with the limitations of the prior art through the incorporation of a new technology.

It refers to machine data security markings that prohibit its unauthorised copying and file identification. The DW-SVD solution for optical video watermarks is recommended [13]. There are two parts: watermark embedding and watermark extraction. In the embedding step, we combine method Comparison with the comprehensive PSNR code tag with the time elapsed and the robustness of the check.

Watermarking approach based on DWT and SVD proposed a hybrid concept of digital video [14]. This method can be used to ensure identification and data privacy. DCT method is rather complicated but it has superior capacity and concealment. It is observed that DWTSVD combination is effective than DCT method. The techniques meet the required security policy, reliability, fast processing and the visibility time required.

Find that steganography works poorly because a very little image shift can be deleted efficiently by rolling LSB from a single pixel is a cover picture stored in JPEG format within the spatial domain [15]. Noted that a large spatial stylus is given in the original uncompressed format, but that uncompressed image sharing is considered equal to a similar author's cryptography.

Suggested a realistic JPEG-embedded image algorithm to deliver steganography with high capabilities without compromising protection [16]. The algorithm F5 is introduced into an irregular coefficient of DCT bit selection and integrated into a matrix to minimize shifts in the length of a post.

A new scheme was proposed to hide LSB replacement for encrypted image data. It consists of the encryption of images, data embedding and data recovery [17]. Encryption key used for encrypting the original image of copyright owners. A shift in the encrypted picture data that masks the use of a key to create a sparse space for the smallest bit of the extra data to handle.

The immune digital watermarking algorithm is the latest, insecure, reverse watermarking the charge [18]. In certain cases reversible digital watermarking will not include the watermark information in a totally restored way to its original undifferentiated state according to the watermark security characteristics, but will allow for minor differentiations. In this sense, this paper suggests an embedded digital watermark that integrates new concepts to guarantee the consistency of work that value has been done after failure, so that the intruder does not benefit from the value of public works embedded with a watermark.

New telemedicine circles are also the solution to enforce image protection requirements by shielding information to safeguard proprietary and copyrighted content [19]. The delicate watermarking system offers a reliable, health-care, telemedicine solution. In this document we suggest a reversible colour medical image, a watermarking method for the spatial domain. Watermarks of an image stored in the unaltered as an important component extract in a given watermarking process. Used to extract the watermark after the main information.

We propose in this paper a hybrid watermarking approach focused on digital images of a single decomposition (SVD)[20]. The watermarks protection of copyright and longevity are two main facets of the software. Here, we entered the wavelet host picture, which provides copyright security and reliability, as main components of DCT field marking. The factor of scaling depends on the image. Providing preliminary findings shows that the software can avoid all forms of attacks and covers in image processing.

Tacit writing is the technology to be covered in some newspapers to prevent eavesdroppers with distrust. They were introducing a new transformation approach for DWT steganography [21]. This approach allows sensitive knowledge to be concealed in a visual image. By changing the wavelet coefficients, hidden knowledge is inserted into the picture. The proposed approach is very similar to the initial hidden picture quality.

Service owners are using forensic watermarking's to track users who upload content unlawfully in order to tackle the redistribution of the piracy technologies in Internet Content-the largest threat to many providers of content around the world [22]. In order to be adaptive, it is necessary that the watermark functions smoothly in this setting, as one of the major frameworks for delivering advanced multimedia content on the internet. In this technical report a solution is defined for a single watermark content with a plurality of scalable streaming applications in each receiving system. The suggested watermarking solutions for service providers include a fully functional and easy-to-use authentication scheme to deter the piracy's redistribution of adaptive broadcasting of internet content.

The authentication, weak or malicious attacks embedded watermark are of significant importance in the strong digital colour image of the RGB colour frame [23]. In this analysis, the most important bit replacement (LS B) for authentication of the RGB colour picture watermarking method is introduced, since the use of a minimized alternate bit (LSB) is based on a distortion of the embedding mechanism. The software is modified in three stages by modifying the initial pixel and the pixel watermark. Initial watermark approaches with red, green or blue histograms for histograms are suggested. However, the 3D image elements are red, green, blue, and watermark too. The initial and watermark image and PSNR algorithm of insertion and removal will be in this article.

The use of digital data is rising with the rapid growth of IT and Multimedia. Thus, it is very critical to secure, but also very difficult, the multimedia material pirating [24]. Many people are afraid that the unauthorized repetition of their knowledge cannot secure their copyright. So, the advancement of technologies is really important in the face of all these issues. Considered the digital option for watermarking to avoid multimedia records. This paper introduces, brings out and incorporates an electronic watermark concept. Under its suggested watermarking, the DWT series was in the original rearrangement format, using a picture which had to be rearranged. Then all bands of high LH, HL and HH were used with DCT and SVD. Then by changing the specific values of these bands, the watermark is integrated.

The watermark is reversed by the optimized watermark extraction process [25]. In order to pick these three strips for a band and a pure high frequency band, strong invisibility is assured and resilient against multiple forms of attacks. Utilising DWT, DCT, and SVD conversion leads to a more efficient algorithm of watermarking than any watermarking algorithm that has been proposed. In terms of hiddenness, watermark picture quality is fine. All bands LH, HL, HH options in this high watermark algorithm are made more resistant for various kinds of filtering noise and geometric noise by the mid-LH, HL and pure high-frequency HH bands. In future, the algorithm can be expanded to colour image and video processing by the entire band DWT-DCT-SVD.

3. Proposed Work

We note that there are a variety of techniques for watermarking a picture using text, images or other means where watermarking is incorporated in the coverage, based on the initiation and inspiration offered by the analysis literature [26]. In addition to many watermark techniques, the least important bit contains (and discrete Wavelet changes). These involve a small bit, a discrete wavelet conversion, a discrete cosine, and several more. Many modern technologies utilise either a single watermark or a different watermark utilising two distinct watermarking techniques. In our analysis we suggest a way of combining two watermarks into a specific video format and evaluating its performance and the quality of its watermarking [27]. Firstly, it allows us in two ways to improve the payload of the cover frame of the film.

3.1 Process of Watermarking

The sender reads a video and extracts its frames, then integrates two optical watermarks, and executes the vectoring procedure for watermarking encryption [28]. In addition, the sender sends the first watermarked initial picture with the LSB technique first plane. Then the consumer combines the cover with the invisible watermark using a 3rd bit plane LSB technique. This fundamental mechanism is replicated across all frames, each with two watermarks [29, 30].

IOP Conf. Series: Materials Science and Engineering 1131 (2021) 012020 doi:10.1088/1757-899X/1131/1/012020

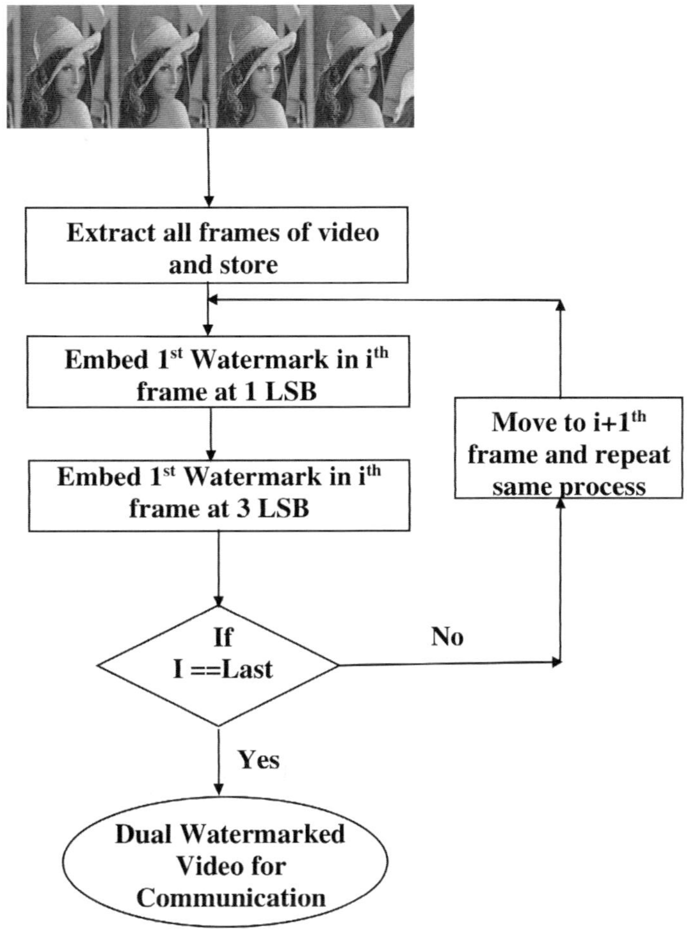

Figure 3. Flowchart at the sender's end

We also used two separate watermarks for embedding:

Watermark 1

Watermark 2

Figure 4. Two different types of Watermark

4. Result & Analysis

The original and watermarked video snapshots are:

1. MP4 (H.264) Video

Original Video Frame (Original) **Watermarked Video with Frame**

Figure 5. Snapshots of Original and Watermarked Video Frame

2. MPEG Video

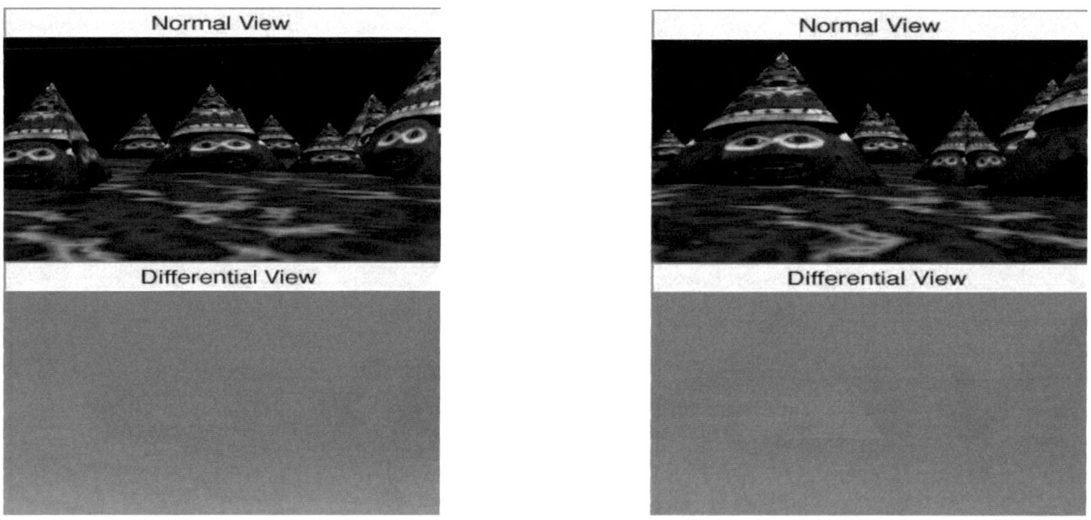

Original Video Frame (original) **Watermarked Video with Frame**

Figure 6. Original and Watermarked Video Frames

It is time to carry out all the tests and interpret the outcomes with the same results. We evaluate the effects of the sender efforts and the results of the marked video.

Please see below the screenshots and effects of the watermarked video of all three forms and their PSNR values:

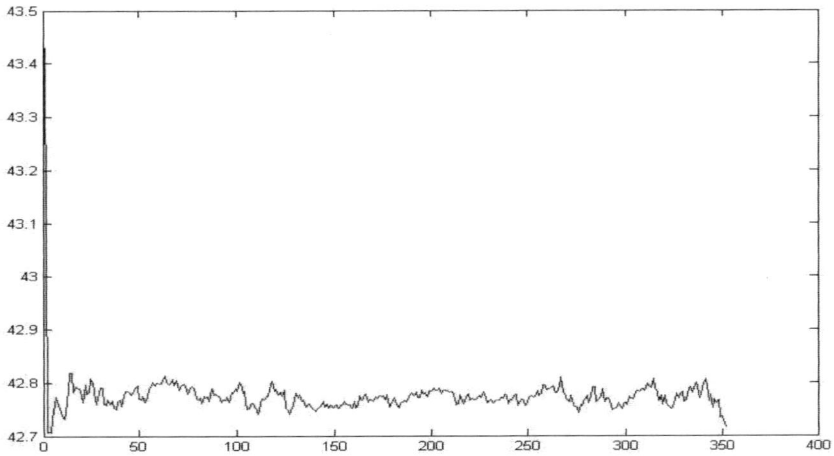

Figure 7. PSNR Graph of all AVI video frames

In addition, the security of a specific frame is strengthened in order to make sure that the video is reliable or right, even though anyone breaks the hidden watermark. This work involves three types of content: MP4, AVI and MPEG footage, and two watermarks are used with each video, enabling us to assess the quality of the watermark in all three formats.

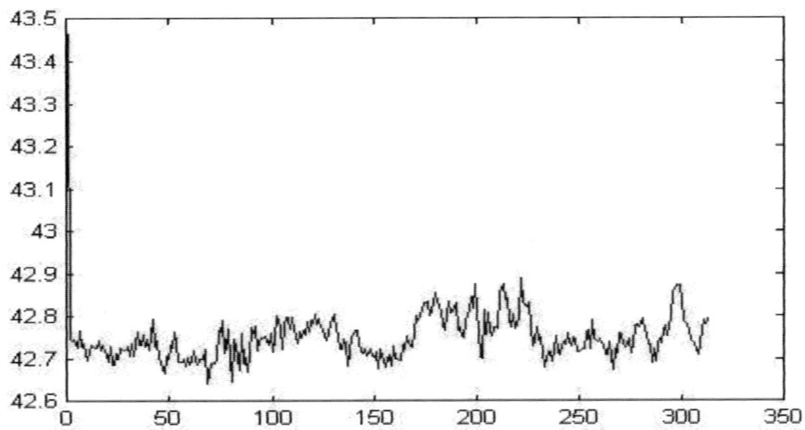

Figure 8. PSNR Graph of all MP4 video frames

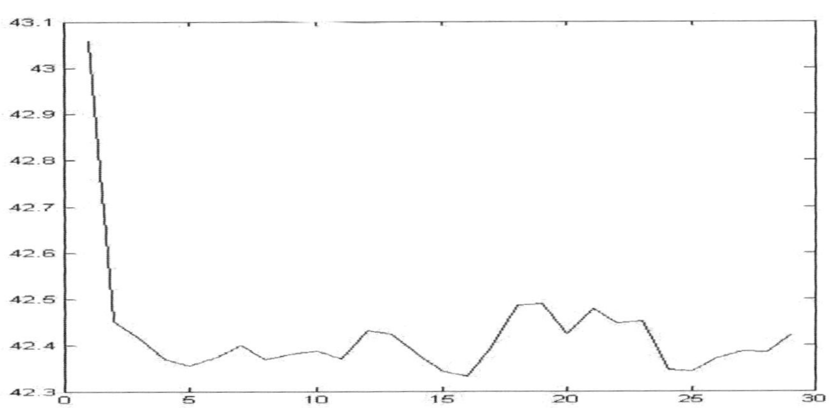

Figure 9. PSNR Graph of all MPEG video frames

5. Conclusion

There are also a range of watermarking techniques that ensure and protect the cover piece, which provides the receiver with limited repetition, with the result of watermarking, like the smallest bit, a discreet wave transformation, discrete cosine transformation, and several more. Many modern technologies utilise either a single watermark or a different watermark utilising two distinct watermarking techniques. In this paper, we suggest that two watermarks of a single video be merged in separate formats and its performance and watermarking quality be evaluated. Firstly, it allows us in two ways to improve the payload of the cover frame of the film. In addition, the security of a specific frame is strengthened in order to make sure that the video is reliable or right, even though anyone breaks the hidden watermark. This work involves three types of content: MP4, AVI and MPEG footage, and two watermarks are used with each video, enabling us to assess the quality of the watermark in all three formats. As our results indicate, with the watermarked video in the input video, the final PSNR performance is very high and the work has also been done to ensure the touch image. The file format H.264 (MP4) and AVI are almost the same. We have also been able to guarantee the authentication and copyright of the watermarked video.

References

[1.] J. Sun, X. Jiang, J. Liu, F. Zhang and C. Li, "An anti-recompression video watermarking algorithm in bitstream domain", in *Tsinghua Science and Technology*, vol. 26, no. 2, pp. 154-162, April 2021. doi: 10.26599/TST.2019.9010050.

[2.] M. Asikuzzaman and M. R. Pickering, "An Overview of Digital Video Watermarking", in *IEEE Transactions on Circuits and Systems for Video Technology*, vol. 28, no. 9, pp. 2131-2153, Sept. 2018.doi: 10.1109/TCSVT.2017.2712162.

[3.] H. Mareen, M. Courteaux, J. De Praeter, M. Asikuzzaman, G. Van Wallendael and P. Lambert, "Rate-Distortion-Preserving Forensic Watermarking Using Quantization Parameter Variation", in *IEEE Access*, vol. 8, pp. 63700-63709, 2020. doi: 10.1109/ACCESS.2020.2984354.

[4.] X. Yu, C. Wang and X. Zhou, "A Hybrid Transforms-Based Robust Video Zero-Watermarking Algorithm for Resisting High Efficiency Video Coding Compression", in *IEEE Access*, vol. 7, pp. 115708-115724, 2019. doi: 10.1109/ACCESS.2019.293613.

[5.] Y. Zhou, C. Wang and X. Zhou, "An Intra-Drift-Free Robust Watermarking Algorithm in High Efficiency Video Coding Compressed Domain", in *IEEE Access*, vol. 7, pp. 132991-133007, 2019. doi: 10.1109/ACCESS.2019.294036.

[6.] H. Mareen, J. De Praeter, G. Van Wallendael and P. Lambert, "A Novel Video Watermarking Approach Based on Implicit Distortions", in *IEEE Transactions on Consumer Electronics*, vol. 64, no. 3, pp. 250-258, Aug. 2018. doi: 10.1109/TCE.2018.2852258.

[7.] X. Li, Y. Wang, Q. Wang, S. Kim and X. Zhou, "Copyright Protection for Holographic Video Using Spatiotemporal Consistent Embedding Strategy", in *IEEE Transactions on Industrial Informatics*, vol. 15, no. 11, pp. 6187-6197, Nov. 2019. doi: 10.1109/TII.2019.2897733.

[8.] Mahrishi M., Morwal S. (2020) "Index Point Detection and Semantic Indexing of Videos—A Comparative Review". In: Pant M., Kumar Sharma T., Arya R., Sahana B., Zolfagharinia H. (eds) Soft Computing: Theories and Applications. Advances in Intelligent Systems and Computing, vol 1154. Springer, Singapore

[9.] M. Amini, M. O. Ahmad and M. N. S. Swamy, "A Robust Multibit Multiplicative Watermark Decoder Using a Vector-Based Hidden Markov Model in Wavelet Domain," in *IEEE Transactions on Circuits and Systems for Video Technology*, vol. 28, no. 2, pp. 402-413, Feb. 2018. doi: 10.1109/TCSVT.2016.260729.

[10.] H. Fang *et al.*, "A Camera Shooting Resilient Watermarking Scheme for Underpainting Documents," in *IEEE Transactions on Circuits and Systems for Video Technology*, vol. 30, no. 11, pp. 4075-4089, Nov. 2020. doi: 10.1109/TCSVT.2019.2953720.

[11.] X. Wang, X. Li and Q. Pei, "Independent Embedding Domain Based Two-Stage Robust Reversible Watermarking," in *IEEE Transactions on Circuits and Systems for Video Technology*, vol. 30, no. 8, pp. 2406-2417, Aug. 2020.doi: 10.1109/TCSVT.2019.2915116.

[12.] Janu, N., Kumar, S., & Mathur, P. (2017). Performance analysis of feature extraction techniques for facial expression recognition. International Journal of Computer Applications, 166(1), 1–3.
[13.] Janu, N., Mathur, P., Gupta, S. K., & Agrwal, S. L. (2017). Performance analysis of frequency domainbased feature extraction techniques for facial expression recognition. 2017 7th International Conference on Cloud Computing, Data Science & Engineering - Confluence. IEEE.
[14.] Neha, & Mathur, P. (2018). Three level optimization models of scaled Gabor features for Facial Expression Recognition. International Journal of Engineering & Technology, 7(2.24), 348.
[15.] Janu, N., & Kumar, A. (2021). Automated fruit grading system using image fusion. In Advances in Environmental Engineering and Green Technologies (pp. 32–45). IGI Global.
[16.] Mr. Pankaj Dadheech, Dr. Dinesh Goyal, Dr. Sumit Srivastava, Mr. Ankit Kumar, (2018), "A Scalable Data Processing Using Hadoop & MapReduce for Big Data", Journal of Advanced Research in Dynamical & Control Systems, Vol. 10, 02-Special Issue, 2018, pp-2099-2109, ISSN: 1943-023X.
[17.] Pankaj Dadheech, Dinesh Goyal, Sumit Srivastava & C. M. Choudhary, (2018), "An Efficient Approach for Big Data Processing Using Spatial Boolean Queries", Journal of Statistics and Management Systems (JSMS), 21:4, 583-591.
[18.] A. Kumar and M. Sinha (2014), "Overview on vehicular ad hoc network and its security issues," International Conference on Computing for Sustainable Global Development (INDIACom), pp. 792-797. doi: 10.1109/IndiaCom.2014.6828071.
[19.] Pankaj Dadheech, Ankit Kumar, Chothmal Choudhary, Mahender Kumar Beniwal, Sanwta Ram Dogiwal & Basant Agarwal (2019), "An Enhanced 4-Way Technique Using Cookies for Robust Authentication Process in Wireless Network", Journal of Statistics and Management Systems, 22:4, 773-782, DOI: 10.1080/09720510.2019.1609557.
[20.] Ankit Kumar, Pankaj Dadheech, Vijander Singh, Linesh Raja & Ramesh C. Poonia (2019), "An Enhanced Quantum Key Distribution Protocol for Security Authentication", Journal of Discrete Mathematical Sciences and Cryptography, 22:4, 499-507, DOI: 10.1080/09720529.2019.1637154.
[21.] Mahrishi, M., Hiran, K. K., Meena, G., & Sharma, P. (2020). Machine Learning and Deep Learning in Real-Time Applications. IGI Global. http://doi:10.4018/978-1-7998-3095-5
[22.] Ankit Kumar, Pankaj Dadheech, Vijander Singh, Ramesh C. Poonia & Linesh Raja (2019), "An Improved Quantum Key Distribution Protocol for Verification", Journal of Discrete Mathematical Sciences and Cryptography, 22:4, 491-498, DOI: 10.1080/09720529.2019.1637153.
[23.] Ankit Kumar and Madhavi Sinha (2019), "Design and analysis of an improved AODV protocol for black hole and flooding attack in vehicular ad-hoc network (VANET)", Journal of Discrete Mathematical Sciences and Cryptography, 22:4, 453-463, DOI: 10.1080/09720529.2019.1637151.
[24.] Ankit Kumar, Vijayakumar Varadarajan, Abhishek Kumar, Pankaj Dadheech, Surendra Singh Choudhary, V.D. Ambeth Kumar, B.K. Panigrahi, Kalyana C. Veluvolu, "Black hole attack detection in vehicular ad-hoc network using secure AODV routing algorithm", Microprocessors and Microsystems, 2020, 103352, ISSN 0141-9331, https://doi.org/10.1016/j.micpro.2020.103352.
[25.] Ankit Kumar, Linesh Raja, Pankaj Dadheech, Manish Bhardwaj (2020), "A Hybrid Cluster Technique for Improving the Efficiency of Colour Image Segmentation", World Review of Entrepreneurship, Management and Sustainable Development, Nov. 2020, Vol. 16, Issue 6, pp. 665-679, Print ISSN: 1746-0573 Online ISSN: 1746-0581, https://doi.org/10.1504/WREMSD.2020.111405.
[26.] Dadheech P., Goyal D., Srivastava S., Kumar A., Bhardwaj M. (2021) Performance Improvement of Heterogeneous Cluster of Big Data Using Query Optimization and MapReduce. In: Goyal D., Bălaş V.E., Mukherjee A., Hugo C. de Albuquerque V., Gupta A.K. (eds) Information Management and Machine Intelligence. ICIMMI 2019. Algorithms for Intelligent Systems. Springer, Singapore, pp 85-100, First Online: 17 September 2020, Print ISBN: 978-981-15-4935-9, Online ISBN: 978-981-15-4936-6. https://doi.org/10.1007/978-981-15-4936-6_9.

[27.] A. Kumar, P. Dadheech and U. Chaudhary, (2020), "Energy Conservation in WSN: A Review of Current Techniques," 2020 3rd International Conference on Emerging Technologies in Computer Engineering: Machine Learning and Internet of Things (ICETCE), Jaipur, India, 2020, pp. 1-8, ISSN: 978-1-7281-1683-9, doi: 10.1109/ICETCE48199.2020.9091736.

[28.] Kumar A., Dadheech P., Beniwal M.K., Agarwal B., Patidar P.K. (2020), A Fuzzy Logic-Based Control System for Detection and Mitigation of Blackhole Attack in Vehicular Ad Hoc Network. In: Chaudhary A., Choudhary C., Gupta M., Lal C., Badal T. (eds) Microservices in Big Data Analytics. Springer, Singapore, First Online: 27 November, 2019, pp 163-178, Print ISBN: 978-981-15-0127-2, Online ISBN: 978-981-15-0128-9, DOI https://doi.org/10.1007/978-981-15-0128-9_15.

[29.] Pankaj Dadheech, Ankit Kumar, Vijander Singh, Linesh Raja, Ramesh Poonia (2020), A Neural Network-Based Approach for Pest Detection and Control in Modern Agriculture Using Internet of Things. In: Amit Kumar Gupta, Dinesh Goyal, Vijander Singh, Harish Sharma (eds), Smart Agricultural Services Using Deep Learning, Big Data, and IoT, October, 2020, pp 1-19, ISBN13: 9781799850038, ISBN10: 179985003X, EISBN13: 9781799850045, DOI: 10.4018/978-1-7998-5003-8.ch001, Publisher IGI Global.

[30.] Ankit Kumar, Pankaj Dadheech, Vijander Singh, Linesh Raja (2020), "Performance Modeling for Secure Migration Processes of Legacy Systems to the Cloud Computing", In: Tin Thein Thwel, G. R. Sinha (eds), "Data Deduplication Approaches: Concepts, Strategies and Challenges", Chapter-13, pp. 255~280, ISBN: 978-0-12-823395-5, DOI: https://doi.org/10.1016/B978-0-12-823395-5.00003-3, Publisher Elsevier.

Comprehensive Analysis of Cloud based Databases

[1]Dr. E. Saravana Kumar, [2]Dr. Selvaraj Kesavan, [3]Dr. R. Ch. A Naidu, [4]Dr. Senthil Kumar R, [5]Latha
[1]Associate Professor,
[2]Advisor Solution Architect in DXC Technology, India,
[3]Professor & H.O.D ,
[4]Associate Profesor,
[5]Assistant Professor
[1, 3,4,5] Dept. of CSE, The Oxford College of Engineering, Bommanahalli, Bangalore-560068
saraninfo@gmail.com, selvarajkesavan@gmail.com, chanr789@gmail.com, senkr.raj@gmail.com, latha.shreeram@gmail.com

Abstract

With standard change in the computing architecture and distributed, wide range processing approach leverage the large data to be analyzed and extract the content. The recent advancement in Web technology enables the user to produce process and ingest content of any structure. Cloud computing helps to provide computing infrastructure, centralized storage and seamless services for organizations and users. In a digital world, data is core aspect of the economy, become part of government, enterprise, social sectors and every individual in a day to day life. Data analysis is gathering of strength in both research and production communities. There are many social networking sites, web, enterprise and IoT data require being stored in secured, centralized data store. It also demands data to be stored without any predefined schema and able to retrieve in near real time. When demand goes up application needs to be scaled to sustain for the demand. Similarly, data store should support scalability to support applications demand. Cloud providers and third-party organization offers numerous data storage, database options. This article highlights various available data storage options, suitability and limitations.

Keywords: Open-Source Database, Cloud storage, Data Storage

1. Introduction

Data science becoming the integral part of every business and individual life. Organizations are spending huge amount in data science to keep up the heat in today"s competition and improve their business outcome. It helps to derive business insights, key metrics and predictive analysis from the data captured. Data science and its relevant fields are highly valuable; it helps management teams to make strategic business decisions. Many companies around the world are exploring the data science and trying to build the data science platforms.

Content from this work may be used under the terms of the Creative Commons Attribution 3.0 licence. Any further distribution of this work must maintain attribution to the author(s) and the title of the work, journal citation and DOI.
Published under licence by IOP Publishing Ltd

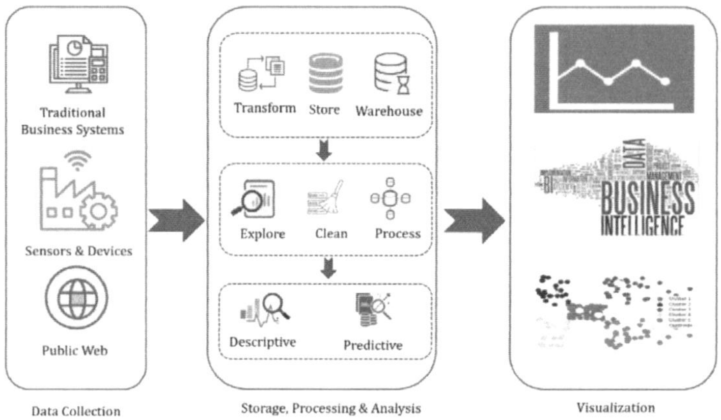

Figure 1. Data science pipeline

With the invent of Web technologies, data with volume, velocity and variety coming from major sources such as Internet of things/sensors, web &social network and traditional business systems. The data source may be structured varieties or unstructured varieties or combinations of both. Data science uses the numerous approaches to clean, explore, transform the data before applying techniques like machine learning deep learning data visualization text analytics to analyze the data. It is tightly integrated with artificial intelligence to make reality of decision-making robots. The Fig.1 shows the architecture of data science Pipeline with detailed manner.

Data science extracts knowledge and provides insights of various form of data for business process efficiency data is important and key part of cloud computing. Data captured from many industry verticals/users such as business & commerce, Entertainment, Science, social network, medical, manufacturing etc.. Data in raw or extracted format brings lot of value to organizations and individual. It helps to expedite the business decisions and improve quality of life. The evolving technologies help to capture Terabytes of data, store in large repositories, and analyze the data with complex data analysis approach using distributed/parallel processing. Data store to Data warehousing helps to store, process, analyze commutative and historical data from single or multiple data sources. Data analysis tools, algorithms and framework helps to derive descriptive and predictive analytics from data and aids to quick decision. The critical aspect of the pipeline is to collect and store the data from diverse sources which produce different data formats.

The data captured from various sources includes structured, single dimension, multidimensional, tabular, spreadsheet, time series, documents formats. The storage mechanism is simultaneously evolving to address the need and support for multiple applications and data formats. The selection of data base is depending on multiple factors such as

- Requirement of Relational between elements
- Amount of data to store
- Real Time Streaming of Data
- Scalability, reliability and availability requirement
- Multi-tenancy
- Cost

Structured Query Language (SQL) is mainly designed for the data requires relational with elements. It deals structured data with defined attributes. The scalability of SQL recognized by Web2.0 companies with large, increasing data and framework needs Facebook, Amazon and Google etc. Because of its sheer size and lack of structure, traditional relational database management system (RDBMS) unable to handle very large big data. These kinds of very large data require a flexible non-relational model that supports fast access to data for processing.

Relational databases are predefined, expensive to scale, and have high storage costs. They are also transactional, consistent, and use mature SQL and a relational mindset. There is no SQL database that can meet this need. On the other hand, non-relational (NoSQL) databases are flexible, scalable, and resilient. They feature commodity storage, are not transaction-oriented, and are eventually consistent. Non-relational databases are also programmable with SQL-like and use a new mindset. No-SQL databases are playing vital role in recent applications in multiple fields. It provides free flow operations without any bottleneck in the performance and use a different types of data models, which includes text document, graphical models, dictionary and columns, helps unique way to query the data. Horizontal scaling and flexible data model definitions. Distributed and developed for very large-scale forms of data storage and enormous parallel processing data across a different number of servers and achieve speed at scale, overcoming RDBMS limitations and inefficiencies. With built-in high-resiliency architecture, it helps to store dynamic schema, semi-structured data for low-latency applications and update, retrieval in real time. This increasing interest among NoSQL and DBMS"s evolved with a focus on improved performance and consistence with reliability. The major No-SQL data bases are Mongo DB(Document), Redis (Key-value),Amazon Redshift (Columnar), Cassandra (Columnar), HBase (Columnar),Dynamo DB(Document DB-stores JSON/XML) and GraphDB .

The major contributions of the proposed paper covers three main points.

- Provides list of various available data storage options, highlights the criteria to choose appropriate database, cache and limitations.
- Elaborately discuss the No-SQL architecture, retrieval mechanism and Caching strategy
- Detailed comparison of various No-SQL storage options with key metrics

Apart from the main points the remaining the paper is organized in other section. Section 2 presents the various related research work in the area of data base and data storage mechanisms and options. The evolution of database paradigm and classifications are discussed in Section 3. The various aspects of No-SQL Database Storage architecture, retrieval mechanism and selection strategy are captured in Section 4. In section 5, the comparison of key No-SQL databases with key metrics are discussed in detail. Finally, section 6 concludes the paper.

2. Related work

Database helps applications to store, update and retrieve information. SQL and No-SQL databases are widely used by the organizations for storing data collected from various sources. Cloud and IoT evolving exponentially, it is critical to have appropriate databases for store, retrieve, and process data on real time. Sensors/Things connected to server via network and deliver connected industry solutions for efficient control and improved human experience. Billions of connected devices are an indicator of IoT and transmits data in high velocity, variety and volume. The real value of IoT is on data which helps business insight and enable data-driven economy.

Database management systems (DBMS) and databases are critical to everyday business and work

[1].Relational Database Management Systems (RDBMS) is old, pioneer system used many years [2]. The need of new databases is inevitable to develop and use new generation digital applications such as social networking, business visualizations, web 3.0 and connected sensors/devices [3]. Authors [4] performs independent investigation on the NoSQL and SQL performance of databases based on dictionary of key-value stores. The various operations such as comparing reading, writing, deleting and instance operations on key-value stores of SQL and NoSQL. Apart from that, authors investigated the operation of iteration across all the keys. A framework of abstract key-value pair [5] designed implemented and tested using the above basic operations. A detailed comprehensive survey [6] of various approaches and deployment mechanisms of data-intensive areas in the cloud gains major attention in both research and industry. An analysis of different design decisions of every approach and its fitness to support various areas of applications and different end-users. Different views of some of the open issues and future questions related to scalability of large scale database in economical processing on the cloud is addressed. Bartholomew [7] explains the differences of SQL and NoSQL databases and its history. The authors [8] compared functional and non-functional features of graph database; dictionary based key value stores, column store database and documents. In each category, authors selected one database on document store (MongoDB), column database (Cassandra) stores, key value stores and graph based databases (Redis and Neo4j)

A qualitative research [9] has been conducted on SQL and No-SQL data base and detailed, intensive analysis and comparison performed. The NoSQL and SQL performance compared experimented and analyzed by the authors [10].IoT and IIoT are one of the major sources in providing data. The authors [11] have analyzed and performed comparisons of SQL, NoSQL in the context of Internet of Things. Authors [12] performed analysis on the effectiveness problem of data format for the purpose of storing and processing in RDBMS technology. SQL and NoSQL databases have been compared [13] to use for an small IoT application of water sprinkler system and analyses the merits, fitness, performance of NoSQL and SQL under different scenarios.

Even though, the various research papers discuss the functionalities and applications of different databases, it does not provide complete view and comparison of multiple databases and applications with various metrics. However, it is important to have all-in-oneview of different databases along with features, architecture, flexibility, limitations and applications. This paper gives comprehensive view of data storage in the market, selection strategy and comparison which helps for developers, organizations to understand the functionalities and choose right data base for their need.

2.1 Evolution of Database and classification

Information Systems and data bases are the two components concurrently existing in the computer science technology. Appropriate data base management system required for storing and retrieval of data in a meaningful manner. RDBMS databases applies the ACID model in transactions management strong atomicity, consistency with Isolation and Durability property. Unfortunately the transaction properties cannot be applied on NOSQL databases because the priorities are different. The NOSQL applies the BASE model which ensures the Basic Availability of the database, Soft state of the model and the Eventual consistency of the database. Over the years, many researchers have conducted various researches in the mentioned field to justify BASE in NOSQL and ACID properties of RDBMS database.

Various structures were applied and improved to enhance the read operation searching techniques. Big companies developed Proprietary NoSQL to focus on their needs. Google developed BigTable and Amazon developed DynamoDB. The successful outcome of these proprietary softwares started the open source community to develop such a software, Hbase, MongoDB , Cassandra, DynamoDB, Hypertable, and Redis are the popular NoSQL database developed by these communities.

2.2 Data quality and governance

2.2.1 Data store

Persistently store and manage repository of data, files, organized and unorganized data. It includes file system, SQL, No-SQL databases etc. Quick access of data on very large number of nodes allowed in non relational database management systems like distributed DBMS. Distributed DBMS has very good access performance and rich query abilities which are limited key-value semantic in other databases. These limitations are noted in Bigtable , Azure and Dynamo models performs comparatively better in peer-to-peer network.

2.2.2 Data Lake

Data lake is a repository to hold large amount of structured, unstructured data accumulated from various sources at any scale. It is act as a raw repository which holds data in original format as it received from the sources. It stores all data types whether structured or unstructured. Storing data in data lake is comparatively cheap .It is suitable for Machine Learning, Predictive analytics, data discovery and profiling

Ex: Amazon S3, Amazon Glacier

2.2.3 Data Warehouse

A data warehouse is a database optimized to store data which are formatted / processed to give shape to the data. The data structure, schema are defined in advance in order to optimize read / write operations. The data helps to derive insights, analytics which helps organization to take management decisions. Storing data in data warehouse is comparatively costlier than Data Lake. It is suitable for Batch reporting, BI and visualizations

Ex: Amazon Redshift

2.2.4 Database

Technology stores data in an organized manner using Data base management system. It is classified as SQL and No- SQL Databases. Entire row is stored Row-oriented databases in a physical block. Secondary indexes used to achieve the high performance in read operations. Instead of packing the whole rows into a block each column organized into set of physical blocks in Column-oriented databases. Better performance achieved in master/slave architecture type with an overhead that single point of failure presented in master node. Every node is identical in peer-to-peer cluster and same responsibilities assigned to each node. Every node achieves fault tolerance with overhead of consistency maintenance is high.

RDBMS, usually executes query language (SQL), is combination interrelated tables. The tables contains rows and columns. The schema represents the relationships among tables and column represents the attribute. These RDBMS databases are highly consistent by design. RDBMS queries are used in banking applications for online transactions. Cloud providers which are popular are IBM DB2 , SQL Server , MySQL and Oracle. MySQL platform is open source cloud platform.

Table model is not used in non-relational databases which is called NoSQL because it stores the content as a document regardless of any structure. Unstructured data is well-suited in non-relational databases technology like image and video contents of social media.

2.2.5 SQL Databases

SQL data bases have predefined schema. Database stores data in table row format in the databases such as MS SQL Server Oracle DB Server open source MySQL, PostgreSQL and SQL databases. Traditionally data warehouse was used in these systems in which online transactional processing (OLTP) implemented better than for analytical purposes.

In a SQL database, all of the columns are read by query for all of the rows to satisfy the query condition It creates a bottleneck in the performance especially analysis related use cases. But in relational database systems it is well suited for transaction processing across many tables . The multiple tables joined using complex join operations.

2.2.6 No-SQL Databases

A different scalable paradigm instead of relational databases are No-SQL data base. The cloud computing currently uses scalable applications but the challenge of scalable database is storing models. NoSQL data base based on cloud can be considered as an alternative. No fixed scheme is required for distributed data bases which used No-Sql. It is well suitable for unstructured data which need to fit into schema based database or the design changes frequently. Different data model varieties utilized such as key- value based data model, columnar data model , graphs and document model

The familiar, widely used No-SQL DB are

- Document -Mongo DB,
- Columnar - Amazon Redshift, Cassandra, HBase, Dynamo DB
- Key-Value – Redis, Riak, Dynamo DB, Berkeley DB
- Graph - Neo4J

2.2.7 Key - Value -Redis, riak

Implementation of NOSQL database is based on Key-value based databases , which are simple . It uses hash method of dictionary . By using hash it matches the key to the values A hash table will be implemented to separately store the unique keys with pointers refers to the corresponding values. Scalar data types can be used for values , for example integer and complex structures. BLOB, List structure and JSON are the complex structures. In NoSQL databases design and implementation can be performed using key value stores. The API such as get, put , delete can be used by clients for reading, writing using NoSQL. These operations are uses primary key concept. It enable scalability and high performance can be achieved.

2.2.8 Document based – Mongo DB, CouchDB

Mongo DB, CouchDB database systems are not relied on schemas which is storing the document. Each record in document databases is an associated data which treats the document as semi-structured data.

A database object encapsulate everything related to document database. It gives less dependency and more agility. The encapsulated and encoded database objects are provided in some formats like JSON an XML . JSON provided object notation of java script.

The JSON , BSON and XML formats uses stored data structure. It uses Key-Value pair in document store and each key has a corresponding value provides the data structure. This is called as *document.* An object containing metadata value with array , string, binary and date. It gives a method to index and apply query based on features in document

2.2.9 Graph based- Neo4j, GraphDB

Graph based databases are entirely different when compared with NOSQL database. Graph based databases that rely on graph structure. In graph structure nodes and edges are connected with others through structure relations like tree. Property Graphs contains *nodes* . The *relationships* among the nodes represented in property graphs. Properties are the attributes which can be store nodes along with relationships. Nodes represented as entities of graph, Node metadata tagged with *labels* used to represent the contexts. Direction provided by relationships which can be bi- directional. The connections are named among two nodes. More than one relationship can be specified between two different nodes and each relationship with a direction, type, start and an end node.

2.2.10 Columnar – Casandra, H-Base

Data stored n Column databases, is cells of columns. These are grouped into families of column. Multidimensional nested sorted map of maps format used in these databases. data identified in the innermost map contains a timestamp. A cell is used to store the data This will be mapped to column and mapped with column family. A Collection of column families found using a row key. The row key on sets of columns implemented for read and write operations. It will be stored as a continuous entry in the storage which is improving the performance.

2.3 Choosing Appropriate Database

Scenario	**Database Type**
Session Storage, user preference, shopping cart data	Key value
Blogging, web Analytics, real time analytics, ecommerce applications	Document Database
Content management, IoT,log analysis	Columnar DB
Spatial Data ,Social network, transport, routing Applications	Graph DB
- Distribute Static and media contents Low cost ,highly durable - Data store for computation and large-scale analytics	Object Storage
Structured data with Query	Relational DB
Meta data store	Key value and Caching
Uninterrupted Key-Value	Object and key value storage

2.3.1 No-SQL Database Storage and retrieval mechanism

The storage and retrieval of No-SQL contains strong consistency. All clients performs two-phase commit protocol on updating the dataset which uses XA transactions, ACID properties. It provides high availability: The availability of requested data to all the clients found at least one copy of the data, even if some of the machines in a cluster is down, Partition-tolerance: the total system keeps its characteristic even when being deployed on different servers, transparent to the client. Flexible schema supported in NoSQL databases which enables developers to quickly adopt the databases. The schema supports structured and unstructured data enables efficient processing which supports high performance in executing queries in the vary large scale environment.

2.3.2 Master less Ring Architecture

Identical role played by all the nodes in master less ring architecture; Without the master node concept all the other nodes are communicating with each other nodes through scalable and distributed protocol called as "gossip."across datacenters architecture based on a peer-to-peer ring can be deployed which can be done by Cassandra.

The Cassandra performs logical division of racks and the required snitches inside the cluster, this installation determines the best node to be stored and rack for replicas of the node

Mongos query routers are used to build the hierarchal architecture MongoDB. The architecture contains one or more routers and one or more shards used to run the mongod. The replica node built the mongodb and it sets consist of a primary node and secondary nodes.

A master-less peer-to-peer architecture architecture along with many data centers is supported by Riak Commercial version architecture. One datacenter acts primary and many different data centers supports with primary node which will be synchronized.

A peer-to-peer architecture Couchbase, consists of cluster manager, index, query in each and every node. The multiple datacenter in Couchbase support, updation flowing from one datacenter to other datacenters. These flowing can be done bilaterally where conflicts can be resolved by each cluster which is the owner of certain set of partitions.

2.3.3 Peer To Peer Architecture Systems

The upshot of the discussion is that the client-server model generates a service hot-spot at the server and increasingly hot network utilization the closer one gets to the server. Traffic that is diffuse toward the client is concentrated near the serve. The system is represented in fig.2

Distributing the service helps. But, as long as clients out number servers, there is a natural imbalance. The solution to this imbalance is a peer to peer architecture where services are provided (hopefully, but not likely) in proportion to their use.

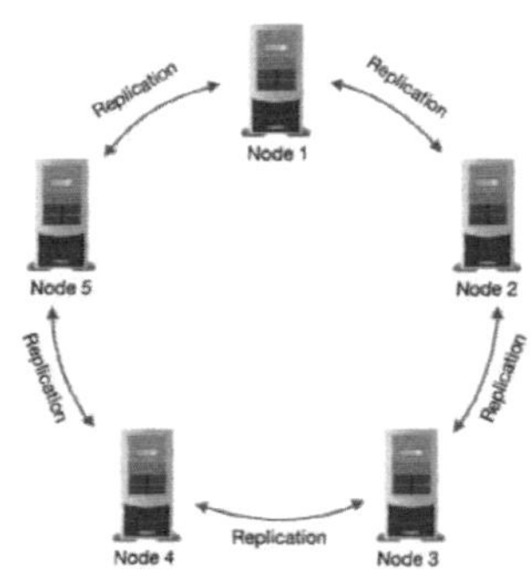

Figure 2: Peer-to-Peer System

3. Challenges of Peer-to-Peer systems

The peer-to-peer architecture systems present several challenges:

- Describing and searching for content
- Naming
- Finding objects and directory services
- Stability of the peers
- Trust of the peers

It becomes difficult to search for object in peer-to-peer systems because high-level searches don't localize well. For this type of thing, we really do want a distributed map-reduce or other parallel search or large, in memory monolithic database, or both. We don't want to have to ask a large number of distant peers and then need to coordinate the results.

Once can identify each object we want, we need to find it. One option is a fully distributed directory service, another is a directory service distributed among select peers, and a third is a distributed hash. Stability of peers is obviously an issue. Some peers will be well resourced and stable, others will be thin and brittle. The stability of the system depends upon having enough stable resources to mitigate the impact of a smaller quantity of brittle resources. One common solution here is to appoint willing, richer, more stable, longer-serving peers as "super peers" and giving them more responsibility, perhaps incentivizing them by proving more or better service (or just by the good feeling from being a good citizen).

Trust is the nearly impossible part. Insert the whole discussion about public key infrastructure here. It is really challenging to trust the identity of hosts. With luck, we can get a hash of what we want from enough sources trusted enough to trust the answer -- and then we can check that what we get matches the hash. Thus, the must brittle part of this is really trusting the search results and/or human-name to hash-name mapping.

3.1 Distributed Hashing: Consistent Hashing and Chords

Another idea for a peer to peer system is to implement a huge distributed hash table. The problem with traditional hash tables, though, is that they don't handle growth well. The hash function produces a large number, which is then taken modulus the table size. As a result, the table size can't change without needing to rehash the entire table -- otherwise the keys can't be found.

A hash value which is independent of the hash table size is used by a hashing scheme in distributed hashing. Keys can be found from the result even if the table size changes. This is not supported of a distributed hash table, The keys found even the nodes enter the system.

Chord protocol is a technique for doing this hashing. A logical ring used to view the world in this protocol. For the selected „m „number of bit key, the bit contains the logical positions 0 to 2m-1. Think of them as hours on a clock. Some of these positions have actual nodes assigned to them, others do not. Every node node "need" the successor only like the token ring, but the topology is known for the entire ring which will be used to handle failures.

For more than one key each and every node is responsible because there are only fewer nodes than actual addresses which is hours on the clock. Mapping is performed on keys to actual nodes by identifying the keys to the "closest" node which can be equal or greater than in number.

a brute force searching technique is use to find a key of the circle, but instead each node keeps a "finger" which points to the next node, 2, or 4 or 8 nodes away, etc. In other words, every node points to nodes exponential manner farther and farther away. These pointers are stored in a table such that the ith entry of the table contains a pointer to a node that is 2i away from it, e.g. at position node number + 2i. The keys, if suppose any node is not present at the exact location then the next passing greater node will be used. This modified arrangement enables the possibility of search for a bucket with the time complexity O(log n) . The time complexity found with each step, which either find the right node, or performs cutting the search space into half.

In order for a node to join, it simply is added to an unrepresented position (hour on the clock) within the hash table. It gets its portion of the keys from its successor, and then goes live. Similarly, disappearing from the hash simply involves spilling ones keys to one's successor.

3.2 Schema less Architecture

NoSQL DB provides Schema less architecture which helps to create schema on-the-fly and it does not need any predefined schema.

Database	Architecture	When to Use
Cassandra	peer-to-peer	**SQL-style data types** **Steep learning curve when switching from SQL**
MangoDB	**master-slave**	**Lots of highly unstructured data** **Loosely coupled objectives**

3.3 Caching Design Patterns and Strategy

- Consistent Caching (Sharding)
    - Issue with Modulo (CRC32) ,keys mapped to new node when scaled
    - Consistent hashing – Ring methodology
    - Client libraries
- Lazy population (updated cache when app reads , cache miss lead to direct query)
- Write on through
    - Cache upgraded to real time when database updated
    - Cache miss avoided
    - Minimize app fetch delay
    - Cache always up-to date
    - Filled unnecessary objects
    - Lot of cache churn due to repeated update
    - No strategy to repopulate cache node when it fails
- Expiration Date
    - TTL to cache keys
    - Short TTL for rapid changing data types (news, leaderboards etc)
    - **Russian doll caching** – nested records with own cache keys , delete only cache keys which needs to be updated (suspect that it will affected by database update)
- Thundering Herd (Dog piling)
    - Cache miss , millions of user hit the DB in parallel increases DB throttling
    - Setting TTL to popular data will impact DB performance (if data not updated but TTL expire will trigger millions to request to be routed to DB)
    - When new cache node added , many requests routed to DB
    - Run script to populate cache before app make requests
    - For new node , run script to populate data in new cache node
    - Prewarming node (incase regular addition/deletion of cache node?
- Cache everything
    - Heavily hit queries and expensive calculations
    - Caching data is stale data (no longer fresh and not appropriate in some scenarios)

3.3.1 Comparison

- HBase (ordered keys, semi-structured data), Cassandra, BigTable,
- CouchDB (name or value in text)
- Sherpa or PNuts (JSON, unordered keys)
- MongoDB (based on JSON)

IOP Conf. Series: Materials Science and Engineering 1131 (2021) 012021 doi:10.1088/1757-899X/1131/1/012021

Name of the tools	**Cassandra Data base model**	**GraphDB**	**MongoDB**	**Neo4j**	**Redis**
Descriptions of the comparative model	Supports wide-column store based on ideas of BigTable and DynamoDB	The graphs uses Enterprise RDF and with reasoning, external index cluster and synchronization support	One of the most popular document stores	The toos supports Open source graph database	**In-memory data structure store, used as database, cache and message broker**
Supported programming languages	Java, JSS,PHP, Py thon,Rub y,S cala,.Net	C#,C++,Clojure, Erl ang,Go,Haskell,J av a, JSS,PHP,Python ,R uby,Scala	Java, JSS,PHP,Pyth on,Ru by,Scala,.Net, coldFu sion	Java, JSS,PHP,P ython,R uby,Scala,. Net,Go, Groovy	**Java, JSS,PHP,Py thon, Ruby,Scala, .Net, Crystal**
The scripting of server-side	no	Java Server Plugin	JavaScript	yes	**Lua**
Partitioning techniques	Sharding supported	none	Sharding supported	none	**Sharding supported**
Replication techniques	Supports selectable replication factor	Supports Master- master replication	Supports Master-slave replication	Causal Clustering using Raft protocol	**Multi-master replication, Master-slave replication**
Consistency concepts of the models	Supports Immediate Consistency , Eventual Consistency	Supports Immediate Consistency, Eventual consistency	Supports Immediate Consistency Eventual Consistency,	Eventual and Causal Consistency configurable in Causal Cluster setup Immediate Consistency in stand-alone mode	**Strong eventual consistency with CRDT**

Transaction techniques	Not supported	ACID Properties supported	ACID Transactions on Multi-document with snapshot isolation	ACID Properties supported	**Atomic execution of commands blocks and scripts and Optimistic locking,**
User concepts supported in models	**Per object the access rights for users can be defined**	**Supported**	**Allows the access rights for users and roles**	**Supports standards wtih Users, roles and permissions. Pluggable authentication (Active Directory, Kerberos, LDAP)**	**Supports password-based access control**

4. Factors impacts database Performance

4.1 Merits

- The databases support elastic scalability which is designed for low-cost commodity hard wares.
- Massive volumes of Big Data Applications can be handled by NoSQL
- Economy: NoSQL databases installation is cheap in commodity hardware even with data volumes and transaction increase. But installation of RDBMS is expensive storage in proprietary servers. Processing and storage is less cost.
- Dynamic schemas: No schema design required for NoSQL databases. In RDBMS, a schema has to be defined first, it makes more difficult because the schema has to be changed every time when the requirements are changed. It ensures that data quality control should be performed on the application. NoSQL has no schema, of the data must be organized properly.
- Auto-sharding: In RDBMS scalability supported vertically, which makes a lot of databases spread among many servers and takes the disk space from multiple servers needed t work. Auto sharding is normally supported in NoSQL databases. Naturally NoSQL automatically spread the data across multiple arbitrary servers. These servers does not required even the application used composition of server pool
- Replication: Automatic database replication is supported in most of the NoSQL databases. This replication used to maintain availability of data all the time. Event maintenance or event of outages maintained using replication. Fully self-healing supported by most of the sophisticated NoSQL which provides automated failover and data recovery. Replication provides the ability to distribution of database across many geographic regions to avoid the regional failures and supports localization.
- Integrated caching: Integrated caching is supported in most of the NoSQL technologies, the frequently-used data kept in the system memory for storing and removing.

4 .2 Limitations of Cloud Database

- Multiple-Data has no relationships among all the data.
- Multi-operation Transactions: Failure to save any one of key among many keys and if any transactions aborted then implementing roll back operation is very difficult
- Query Data by 'value': It support the searching by keys, which is based on information found in the 'value' of the key.
- Operation by groups: Simultaneously several keys cannot run as operations are grouped to one key at a time

5. Conclusion

If the application consists of a small amount of traffic Cloud Datastore will work well and performs better. if database is scaled up to very large number of requests then performance decreases. No wasted capacity will be made for the request given. It is expensive than using data store and required to know the capacity in advance. These will be charged for the additional capacity. Amazon announced auto scaling will be supported in cloud database in near future. Getting the provisioned capacity will help to write the own scripts to handle the scalability issue. Depending on the more capacity requirement it is provisioned, then start to deplete accumulated credits. If credits depleted, then the request will be throttled or failed some time. Depending on the options, the scalability available databases is alluring. The tradeoffs and selecting the right pricing model suitable for the application, enable to free the user from constraints of RDBMS.

References:

1. K. L. Berg, T. Seymour and R. Goel, "History of databases," International Journal of Management & Information Systems, vol. 17, no. 1, pp. 29-35, 2013.
2. U. Bhat and S. Jadhav, "Moving Towards Non-Relational Databases," International Journal of Computer Application, vol. 1, no. 13, pp. 40-46, 2010.
3. R. P. Padhy, M. R. Patra and S. C. Satapathy, "RDBMS to NoSQL: Reviewing Some Next-Generation," International Journal of Advanced Engineering Science and Technologies, vol. 11, no. 1, pp. 15-30, 2011
4. Y. Li and S. Manoharan, "A performance comparison of SQL and NoSQL databases," *2013 IEEE Pacific Rim Conference on Communications, Computers and Signal Processing (PACRIM)*, Victoria, BC, 2013, pp. 15-19.
5. J. C. Anderson, J. Lehnardt, and N. Slater, CouchDB: The Definitive Guide. O'Reilly Media, January 2010.
6. S. Sakr, A. Liu, D. Batista, and M. Alomari, "A survey of large scale data management approaches in cloud environments," Communications Surveys Tutorials, IEEE, Vol. 13, no. 3, pp. 311-336, 2011.
7. Mahrishi M., Shrotriya A., Sharma D (2012). "Globally Recorded binary encoded Domain Compression algorithm in Column Oriented Databases" in Global Journal of Computer Science and Technology, [S.l.], ISSN 0975-4172.
8. A. Gupta, S. Tyagi, N. Panwar, S. Sachdeva and U. Saxena, "NoSQL databases: Critical analysis and comparison," *2017 International Conference on Computing and Communication Technologies for Smart Nation (IC3TSN)*, Gurgaon, 2017, pp. 293-299, doi: 10.1109/IC3TSN.2017.8284494.
9. K. Sahatqija, J. Ajdari, X. Zenuni, B. Raufi and F. Ismaili, "Comparison between relational and NOSQL databases," *2018 41st International Convention on Information and Communication Technology, Electronics and Microelectronics (MIPRO)*, Opatija, 2018, pp. 0216-0221, doi: 10.23919/MIPRO.2018.8400041.
10. Kumar, A., Vengatesan, K., Vincent, R., Rajesh, M., Singhal, A. : A novel Arp approach for cloud resource management; International Journal of Recent Technology and Engineering (IJRTE) at Volume-7 Issue-6, March 2019
11. K. Fraczek and M. Plechawska-Wojcik, "Comparative Analysis of Relational and Non-relational

Databases in the Context of Performance in Web Applications", *Proceedings 13th International Conference Beyond Databases Architectures and Structures (BDAS 2017)*, pp. 153-164, May 30 - June 2, 2017.

12. F. Haleemunnisa and W. Kumud, "Comparison of SQL NoSQL and NewSQL Databases for Internet of Things", *IEEE Bombay Section Symposium*, pp. 1-6, Dec. 2016.
13. K.B. Kumar, SrivydiaSundhara and S. Mohanavalli, "A performance comparison of document oriented NoSQL databases", *Computer Communication and Signal Processing (ICCCSP) 2017 International Conference on*, 2017.

IOP Conf. Series: Materials Science and Engineering 1131 (2021) 012022 doi:10.1088/1757-899X/1131/1/012022

An adaptive approach for word sense disambiguation for Hindi language

Vimal Kumar Soni[†], Dinesh Gopalani[‡] and M C Govil[§]

Department of Computer Science and Engineering

Malaviya National Institute of Technology, Jaipur, India

E-mail: [†]ruvimals@gmail.com, [‡]dgopalani.cse@mnit.ac.in, [§]govilmc@gmail.com

Abstract. Word sense disambiguation has been a classical problem and it is still considered as a big challenge for NLP researchers. Word vectors have shown remarkable improvements from baseline performances for various NLP challenges. An adaptive approach is developed to resolve the ambiguity at word level using word vectors and presented in this paper. CBOW and Skip-gram models of Word2Vec are used to generate the word embeddings. An input sentence is split into two sets: The first set contains all possible ambiguous words and the second set is constituted from all the unambiguous words. The words from the first set are processed through IndoWordNet to collect all the senses of respective words, and such senses are stored in sense-dictionary. An adaptive approach is applied to find out the exact interpretation of the ambiguous word from the sense dictionary for each word from the set of ambiguous words with the context words from the set of unambiguous words. The proposed approach is evaluated on a large scale Hindi corpus and claimed better results than previous attempts.

1. Introduction

Natural Languages are ambiguous by default. A sentence can be ambiguous due to one or many ambiguous words or due to the order of the words. Sometimes the punctuation also plays a vital role in multiple meaning of a sentence. Word Sense Disambiguation (WSD) is used to find out the correct and relevant meaning (sense) of a word which can have multiple meaning in the sentence. Without applying Word Sense Disambiguation, a sentence can be interpreted in more than one meaning. Out of these multiple meanings of the sentence, at least one of the meanings will be always correct however sometimes more than one meaning are also sensible. The WSD is counted in initial steps of language processing, as without it, sense of a sentence can be misinterpreted [1].

In normal human communication, humans perceive secondary information like context from the environment and the sender of the information. With the support of such information, humans can perceive the right meaning of a word out of multiple meanings easily. However such crucial information is not available in the text while processing it by machines which results to develop the correct meaning of the sentence is a challenging task.

WSD is defined as the "ability to identify the meaning of words in context in a computational manner" by Navigli[3]. A word may act ambiguously at any place in a sentence or the word may act ambiguously in a specific role or position in the sentence. In the first case, it is required to perform WSD on all words in a sentence however in the later one WSD is applied only on the specific word. Based on this, there are two classifications- WSD for all words in a sentence and WSD for Lexical-sample.

Researchers are taking advantage of huge web crawled data in the form of corpus and applying different deep/shallow learning methods to generate word embeddings or word vectors. A word embedding or

Content from this work may be used under the terms of the Creative Commons Attribution 3.0 licence. Any further distribution of this work must maintain attribution to the author(s) and the title of the work, journal citation and DOI.
Published under licence by IOP Publishing Ltd

word vector is a learned representation for words where the words with similar semantics have related and close representation.

In humans, the meaning of a word is associated with certain neurons and weights among their interconnection in the brain. Similar words with related meaning form logical groups associated together. The word vectors also follow the same principle. The dimensions of such vectors vary from 50-600 as per the requirement of the application and availability of computing resources. These word vectors are represented using real numbers to imitate the behaviour of a word and have performed extremely well in various NLP tasks. There are various approaches to generate word embeddings like Hard and Soft Clustering models, Mathematics based counting models and Neural-Network based models.

The word vectors generated using neural-network based models specifically word2vec and fastText outperformed vectors generated by other models in most of the NLP tasks. Most of the research works have been carried out in the English language for the generation of word vectors for the English Language. Researchers have focused more on certain languages like Arabic, Chinese, French Japanese and Russian. However, few attempts like Polyglot have been made to develop word vectors in a group for multiple languages in one go. For the Hindi Language too, continuous efforts are in progress to develop quality word vectors. This paper is an attempt to create word embeddings using word2vec and use these word embeddings for word sense disambiguation.

The next section presents an extensive survey of the work done in the domains of WSD and Word Representation respectively. Section III describes the proposed work and after that, the methodology is explained in Section IV. Experimental results are discussed in Section V. The last section covers the conclusion and future work in a nutshell.

2. Related Work

2.1. Word Sense Disambiguation:

The automatic WSD is considered as a classical problem in NLP domain from the starting of automated language processing. With the genesis of automatic machine translation, WSD was assumed as a sub-task. Numerous research attempts have been made for the English WSD, however, for the Hindi language, WSD challenge is still a hot topic.

Ide et al. discussed WSD in detail and termed it as "AI-Complete". That is, "a problem which can be solved only by first resolving all the difficult problems in artificial intelligence (AI), such as the representation of common sense and encyclopedic knowledge" [1]. Agirre and Stevenson [2] provided a thorough analysis of finding clues through the syntactic and semantic analysis of a sentence in 2007. Navigli relied on knowledge-based approaches and also mentioned that supervised methods outperform unsupervised methods [3]. Zhong et al. took designed a system to take leverage of POS tagging, context window of 5 surrounding words and local collocations to solve the challenge of WSD [4].

Taghipour et al. proposed a framework using word embeddings and previous model with changing the context window to 7 surrounding words and 11 local collocations around the word [5]. Rothe and Hinrich used synsets and word embeddings to propose a model with an assumption that a word was a sum of its lexeme [6]. Chen et al. presented a unified model which developed one representation per sense and outperformed other systems [7]. They also developed domain-specific embeddings focusing on Sports and Finance domain.

Research Work in for WSD in the Hindi language has not been as vast as it is in the English Language. The first attempt is considered by Sinha et al. who applied a statistical technique to disambiguate words with the help of wordnet [8] in 2004. Mishra et al. proposed a solution using a classifier which learned from unlabeled Hindi corpus [9] in 2009. Singh et al. proposed an algorithm using hypernym, hyponym, holonym and meronym relations to solve the ambiguity and proved that hyponyms could help.[10]. Agarwal and Bajpai et al. explored the disambiguation of nouns using wordnet in 2014 [11].

Patel and Bhatt proposed a solution based on Hierarchical clustering technique [12]. In 2015, Tayal et al. presented a solution using Fuzzy C-Means Clustering to create sense topics with which they resolved the ambiguity of the word [13]. In 2016, Singh et al. take a step forward in the direction to resolve ambiguity by developing annotated sense corpus [14]. In 2018, Athaiya et al. presented a genetic new approach using genetic algorithm with the feature of a dynamic window subject to the condition of availability of topic in the neighbourhood of the ambiguous word [15]. Sharma and Joshi developed a system which used the knowledge-based LESK algorithm and Wordnet [16]. Kumari and Lobiyal proposed a solution to use word embeddings to resolve the ambiguity [17].

2.2. Distributed Representation of Words

The one-hot representation is one where each word is represented as a sparse vector of the same size of the vocabulary and only one value is set 1. The method of representing a word in one-hot vector representation suffers from the curse of dimensionality as the one-hot vectors are quite large yet sparse. Distributed representation of words (Word Embeddings) resolved this challenge by representing the vectors in limited dimensions with real values. Consider a vocabulary created using the following two documents: D1 and D2:

D1: यह आम रास्ता नही है|

D2: मुझे आम खाना पसन्द है|

So after removing stop-words, the resultant Vocabulary V is.

V = { 'आम', 'रास्ता', "खाना" , "पसन्द" }

Examples of one-hot representation and distributed representation are mentioned in Table-1 and Table-2 respectively.

Table 1. One-hot representation of words

Word	Word Vector			
आम	1	0	0	0
रास्ता	0	1	0	0
खाना	0	0	1	0
पसन्द	0	0	0	1

Table 2. Distributed representation of words

Word	Word Vector
आम	{0.7766, -0.3528, 0.1335, -0.3631, -0.5277}
रास्ता	{0.1507, 0.0229, -0.2650, -0.3493, 0.4425}
खाना	{0.647, 0.2823, -0.3315, 0.6313, -0.7752}
पसन्द	{0.3071, -0.2923, 0.5023, 0.9334, -0.5414}

In 1986, Rumelhart et al. introduced the concept of backpropagation [18]. Bengio et al. introduced word embeddings in 2003 with their neural probabilistic language model [19]. This model was developed using a feed-forward neural network to forecast the next word. Collobert and Weston fetch semantic information from a large corpus in the form of word embeddings in their unified architecture [20]. Mikolov et al. in 2013 brought a kind of revolution in the research world by proposing CBOW and Skip-gram Models and word2vec tool [21].

These two models did not only generate effective word vectors on the basis of the semantics of the word but also were able to manipulate the meaning of the word if its vector is changed after performing vector operations on it. The CBOW takes surrounding words and forecasts the target word, on the contrary, the Skip-gram model accepts a word and predicts its neighbour/context words. Joulin et al. used sub-word information like morphological variations to generate improved vectors [23]. Different Embedding Models and their objective functions are presented and diccussed in detail in [24].

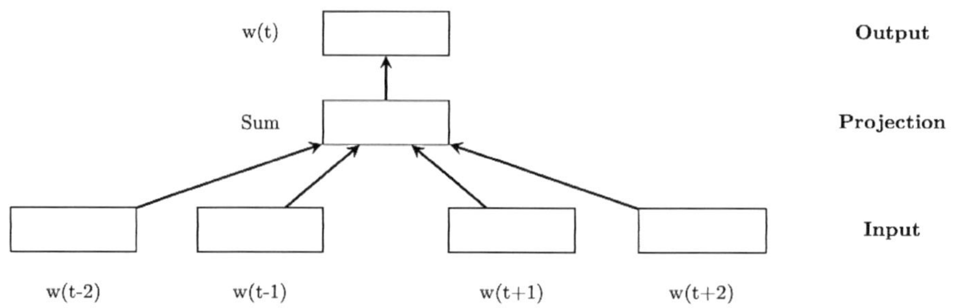

Figure 1. Continuous bag-of-words (CBOW) Model [21]

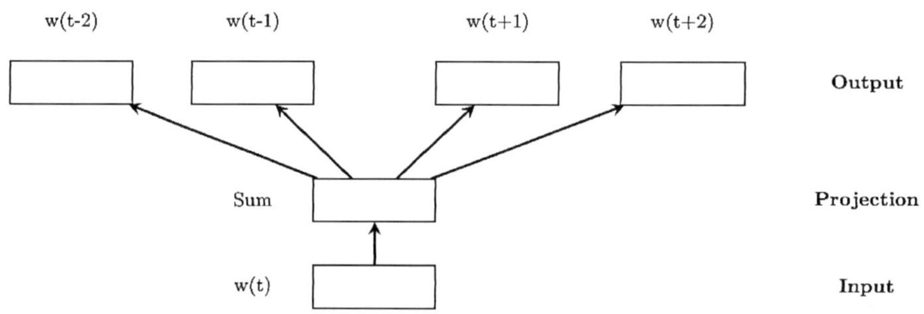

Figure 2. Skip-Gram Model [21]

In this work, we have used variations in tuning parameters of word2vec. Three tuning parameters are considered for such variations are– 1. Vector Dimensions, 2. Window Size and 3. Negative Sampling.

3. Proposed Approach

First, the corpus is collected using various methods and then it was pre-processed using standard methods to clean and make it ready to use. This pre-processing includes Tokenization and Normalization with the following steps:

- Remove numbers and special characters
- Replacing punctuation with white spaces
- Stripping white spaces
- Filtering stop-words

Siddiqi et al. have presented an extensive list of more than 800 stop-words in their research work [25]. This list has been used in this research work to filter stop-words from the corpus as well as from the test sentence. Afterwards, the word vectors are generated using this pre-process corpus. These word vectors are later used in the second algorithm to map words from the given test sentence to their vector forms. The first algorithm WSD_VectorGeneration is shown below:

Algorithm : WSD_VectorGeneration (Corpus, D, W, NS)

Input: Corpus, Tuning Parameter
Output: Word Vector Repository

1. Preparation of Corpus
2. Pre-processing of corpus
3. Set tuning parameters
4. Generate Word Vectors and store in Vector Repository

Algorithm 1: WSD_VectorGeneration

The Second Algorithm WSD_VectorProcessing written inline with Python language is shown below.

```
Algorithm : WSD_VectorProcessing (Vector_Dictionary)
Input: Vector_Dictionary, Test_Sentence
Output: The Best One Sense out of multiple meanings
        for the ambiguous word which suits the best
        semantically in the Test Sentence

1.  Input Test Sentence
2.  Pre-processing and cleaning of Test Sentence
3.  Tokenization to create word list WL
4.  for each word in WL do:
        if ambiguous(word)==True:
            AS.append(word)
        else
            US.append(word)
        end if
    end for
5.  Create Empty Context Dictionary CD
6.  for each word in AS do:
        context = US + AS - word
        CD[word]= [context]
    end for
7.  Create Context Vectors Dictionary CVD
8.  for each word in CD do:
        temp_vec=0
        for each contextword in CD[word] do:
            temp_vec=temp_vec+word2vec(contextword)
        CVD[word]= temp_vec / len(CD[word])
        end for
    end for
9.  Create Empty Sense Dictionary SD
10. for each word in AS do:
        senses = IndoWordNet_FetchSense(word)
        SD[word]= [senses]
    end for
11. Create Sense Vectors Dictionary SVD
12. for each word in SVD do:
        sense_vec=[]
        for each sense in SD[word] do:
            temp_vec=0
            for each part in sense do:
                sense_vec.append=word2vec(part)
            end for
        SVD[word] = sense_vec
        end for
    end for
13. Create Empty Reference Vector Dictionary RVD
14. for each word in SVD do:
        temp=id=0
            for each vector in SVD[word] do:
                temp= CVD[word]-(sum(SVD[word])-vector)
                RVD[[word,id]]= vector + temp
            end for
    end for
15. Create Test Sentence Vector TSV
16. Generate Sentence Decode Vectors' Matrix SDV
17. Choose vector from SDV such that it has minimum
    cosine distance from TSV
18. The id represents the index of the word with
    the best applicable sense w.r.t. Input Sentence
```

Algorithm 2: WSD_ VectorProcessing

A given test sentence is segregated to two sets: The first set contains all possible ambiguous words from the test sentence and the second set is constituted from all the unambiguous words. The words from the first set are processed through IndoWordNet to collect all the senses of respective words, and such senses are stored in sense-dictionary. An adaptive approach is applied to discover the exact sense of the ambiguous word from the sense dictionary for each word from the set of ambiguous words with the help of context words from the set of unambiguous words. Finally, the Cosine Similarity is used as a standard measure to evaluate the semantic relation between the word vectors.

4. Methodology

This system works in two phases. In the first phase, the corpus is pre-processed and Word Vectors are generated from this pre-processed corpus. The corpus used to implement this paper is formed using two different parts: first, Indian Corpora from NLTK and second, Web crawled data from Wikipedia. After forming the corpus by amalgam, it was pre-processed using the steps mentioned in the previous section. To generate word vectors from this corpus, we have to finalize the tuning parameters of CBOW and Skip-gram models.

The variations in tuning parameters are arranged in three different combinations for both CBOW and Skip-gram. In the first combination, the dimensions of vectors are fixed as 50, with a window size of 15 words and no negative sampling. The second combination consists of the vector dimensions as 150, window size as 10 words and with a negative sampling of 5 words. The vector dimensions and negative sampling size are increased to 200 and 10 respectively while the window size is decreased to 5 in the last combination. 100 and 150. With this setup, the word vectors are generated. The same process is shown in Figure-3.

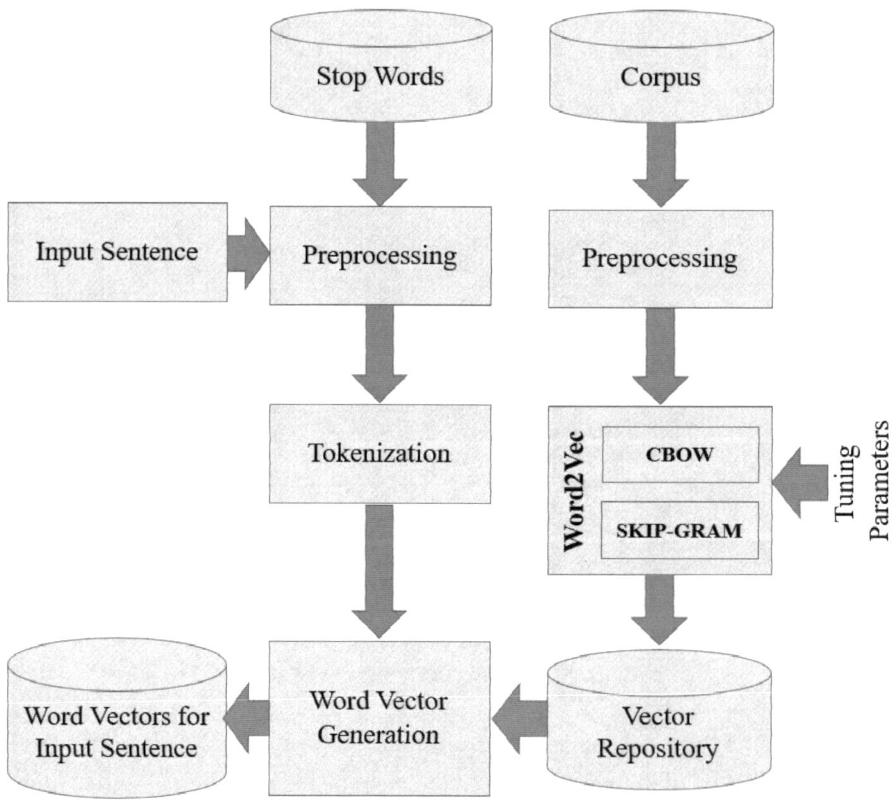

Figure 3. Generation of word vectors

The steps of the first algorithm are self-explanatory, however, the second algorithm is explained below due to its complexity:

1. Input Test Sentence
2. Preprocessing and cleaning of Test Sentence
3. Tokenization
4. Segregation of Words to Ambiguous and non-ambiguous word sets

$$AS = \{w_1, w_2, \dots, w_A\}$$

Where AS represents the Ambiguous Word Set, and
A denotes the total number of ambiguous words in the set

$$US = \{w_1, w_2, \dots, w_U\}$$

Where US represents the Unambiguous Word Set, and
U denotes the total number of ambiguous words in the set

5. Construction of Context Dictionary from the Set of unambiguous words with respect to each ambiguous word from Ambiguous Word Set

$$CD = \{w_1 : [w_{11}, w_{12}, \dots, w_{1p}], w_2 : [w_{21}, w_{22}, \dots, w_{2q}], \quad \dots w_A : [w_{A1}, w_{A2}, \dots, w_{Ar}]\}$$

Where CD represents the Context Dictionary.
p represents the total number of context words for word-1
q denotes the total number of context words for word-2
r symbolizes the total number of context words for word-A

6. Construction of Context Vectors Dictionary from Context Dictionary with a 1 to 1 mapping of Ambiguous Word to its Context Vector.

$$CVD = \{w_1 : CV_1, w_2 : CV_2, \dots \quad w_A : CV_A, \}$$

Where CVD represents the Context Dictionary.
CV1 represents the context vectors for word-1 and
CVA represents the context vectors for word-A
And A represents the total number of words in the dictionary.

7. Fetch sense of each word of the ambiguous set using Indowordnet. The Sense Dictionary is then updated by entering the word and its fetched (multiple) meanings from the Indowordnet.

Since Ambiguous word set has a total of "A" words so the Sense dictionary will also have total "A" entries. However, there will be multiple entries of senses against each word.

$$SD = \{w_1 : [S_{11}, S_{12}, \dots, S_{1X}], w_2 : [S_{21}, S_{22}, \dots, S_{2Y}], \dots w_A : [S_{A1}, S_{A2}, \dots, S_{AZ}]\}$$

Where SD represents the Sense Dictionary,
A is the total number of words in the SD
S_{1X} represents the Xth Sense for word-1 and
S_{AZ} represents the Zth Sense for word-A.

8. Construction of Sense Vectors Dictionary for each sense/ meaning for each Ambiguous Word with the help of Sense Dictionary

$$SVD = \{w_1 : [SV_{11}, SV_{12}, \dots, SV_{1X}], w_2 : [SV_{21}, SV_{22}, \dots, SV_{2Y}] \\ \dots w_A : [SV_{A1}, SV_{A2}, \dots, SV_{AZ}]\}$$

Where SVD represents the Sense-Vector Dictionary.
SV_{1X} represents the sense vector of word-1 for X^{th} sense
SV_{AZ} represents the sense vector of word-A for Z^{th} sense

9. Calculation of Reference Vectors with the help of Context Vector and Sense Vector using following equation-

$$R_{ij} = SV_{ij} + \sum_{k=1,k\neq j}^{n_i} CV_i - SV_{ik}$$

Where R_{ij} denotes reference vector of the ith word for j^{th} Sense,
SV_{ij} represents the Sense Vector of ith word for j^{th} Sense
CV_i refers to the Context Vector of ith word, and
n_i symbolizes the total number of senses for i^{th} word

10. Generation of Test Sentence Vector with all words

$$TSV = V_{w_1} + V_{w_2} + \cdots + V_{w_m}$$

Where TSV is a composite vector of The Test Sentence, and
V_{w_1} represents the Vector of word1 and
V_{w_m} denotes the vector of word m.
where m is the length of Sentence in terms of words.

11. Generation of Sentence Decode Vectors- It is the same as generating TSV except that to create this, we put the Reference Vector instead of Vectors of ambiguous words. Since ambiguous words will have more than one meaning so this will result in a matrix of multiple vectors. This SDV Matrix, each column represents one word and one row represents a complete sentence with one meaning of the ambiguous word and thus this SDV matrix contains all the possible interpretations of the sentence which might be meaningful or meaningless.

$$SDV_{ij} = V_{w_1} + R_{ij} + \cdots + V_{w_m}$$

Where m is the number of words in Test Sentence after tokenization.

12. Selection of the reference vector which is semantically most close to Vector of target Sentence can be done using

$$Sim(\,TSV, SDV_{ij}) = argmax\, sim(TSV, SDV_{ij})$$

$$Sim(\,TSV, SDV_{ij}) = argmax\, \frac{TSV.\ SDV_{ij}}{\|TSV\|.\|\,SDV_{ij}\|}$$

13. The sense related to the vector (SDV_{ij}) with a minimum cosine distance from TSV is the best interpretation of the ambiguous word.

To explain the process further, we have taken a test sentence with an ambiguous word – “आम”.

Test Sentence 1:

बड़े शहरों में तो यह समस्या आम इंसान की है |

After removing stop words and cleaning:

बड़े शहरों समस्या आम इंसान

Context Vectors:

For each ambiguous word, we need to create one context vector.
For Test Sentence 1:

CV_1 = (V(बड़े) + V(शहरों)+ V(समस्या) + v(इंसान))

The Glosses of the target word "आम", fetched from IndoWordNet are shown in Table-3.

Table 3. Glosses of word - "आम" from IndoWordNet

Gloss No.	Synset ID	Gloss	POS
Gloss1	3462	एक फल जो खाया या चूसा जाता है	noun
Gloss2	3468	जिसमें कोई विशेषता न हो या अच्छे से कुछ हल्के दरज़े का	adjective
Gloss3	3463	गर्म देशों में पाया जाने वाला एक बड़ा, सदाबहार पेड़ जिसके रसीले फल खाए या चूसे जाते हैं	noun
Gloss4	3469	प्रायः सभी व्यक्तियों, अवसरों, अवस्थाओं आदि में पाया जाने वाला या उनसे संबंध रखने वाला	adjective
Gloss5	2283	जो पका हुआ न हो	adjective
Gloss6	2697	जो कम पका हुआ हो	adjective
Gloss7	6253	एक रोग जिसमें शरीर से चिकना, सफ़ेद, लसदार मल बार-बार निकलता है	noun
Gloss8	8466	एक प्रकार का लसदार चिकना सफेद मल जो अन्न न पचने से उत्पन्न होता है	noun
Gloss9[a]	Blank	Blank	adjective

[a] 9 Synset fetched from IndoWordNet for "आम" including 1 blank Synset.

The process of WSD_ VectorProcessing() up to the creation of Reference Vector is depicted in Figure4.

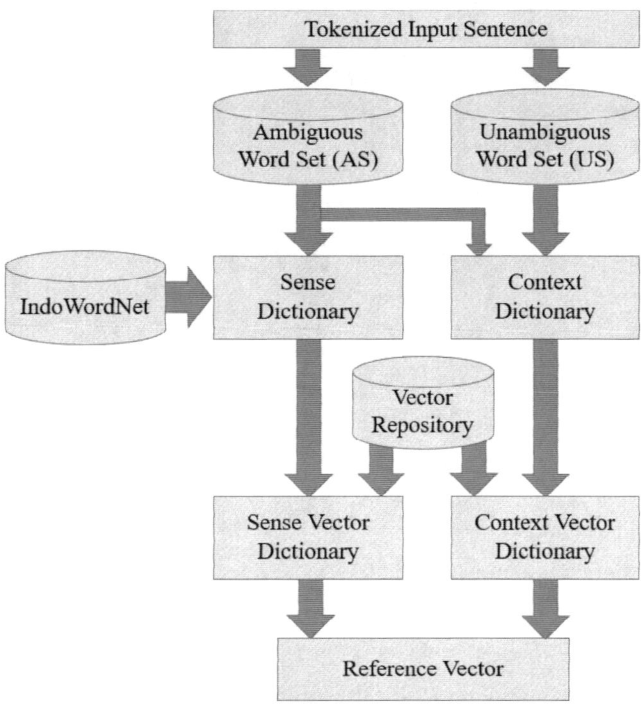

Figure 4. Generation of Reference Vector by Vector Processing

Sense Vectors:

$SV_{\text{आम1}}$ = (V(फल) + V(खाया)+ V(चूसा))/3

$SV_{\text{आम2}}$ = (V(विशेषता) + V(अच्छे) + V(हल्के) + V(दरज़े))4

$SV_{\text{आम3}}$ = (V(गर्म) + V(देशों) + V(पाया) + V(बड़ा) + V(सदाबहार) + V(पेड़) + V(रसीले) + V(फल) + V(खाए)+V(चूसे))/10

$SV_{\text{आम4}}$ = (V(प्राय) + V(व्यक्तियों) + V(अवसरों) + V(अवस्थाओं) + V(पाया) + V(संबंध))/6

$SV_{\text{आम5}}$ = (V(पका))/1

$SV_{\text{आम6}}$ = (V(कम) + V(पका))/2

$SV_{\text{आम7}}$ = (V(रोग) + V(शरीर) + V(चिकना) + V(सफ़ेद) + V(लसदार) + V(मल) + V(बार) + V(चिकना) + V(निकलता))/9

$SV_{\text{आम8}}$ = (V(लसदार) + V(चिकना) + V(सफ़ेद) + V(मल) + V(अन्न) + V(पचने) + V(उत्पन्न))/7

$SV_{\text{आम9}}$ = Blank

Reference Vectors:

Reference Vectors are to be created for each sense for an ambiguous word. There are 9 Synsets mentioned for the ambiguous word – "आम", however only 8 glosses/senses are available, therefore only 8 reference vectors are generated.

$$R_{\text{आम}1} = SV_{\text{आम}1} + \sum_{k=1,k\neq 1}^{8} CV_1 - SV_{\text{आम}k}$$

$$R_{\text{आम}2} = SV_{\text{आम}2} + \sum_{k=1,k\neq 2}^{8} CV_1 - SV_{\text{आम}k}$$

$$R_{\text{आम}3} = SV_{\text{आम}3} + \sum_{k=1,k\neq 3}^{8} CV_1 - SV_{\text{आम}k}$$

$$R_{\text{आम}4} = SV_{\text{आम}4} + \sum_{k=1,k\neq 4}^{8} CV_1 - SV_{\text{आम}k}$$

$$R_{\text{आम}5} = SV_{\text{आम}5} + \sum_{k=1,k\neq 5}^{8} CV_1 - SV_{\text{आम}k}$$

$$R_{\text{आम}6} = SV_{\text{आम}6} + \sum_{k=1,k\neq 6}^{8} CV_1 - SV_{\text{आम}k}$$

$$R_{\text{आम}7} = SV_{\text{आम}7} + \sum_{k=1,k\neq 7}^{8} CV_1 - SV_{\text{आम}k}$$

$$R_{\text{आम}8} = SV_{\text{आम}8} + \sum_{k=1,k\neq 8}^{8} CV_1 - SV_{\text{आम}k}$$

Now after obtaining these Reference Vectors for the ambiguous word, the Sentence Decode Vectors are computed as mentioned below.

$$SDV_{\text{आम}1} = V_{\text{बड़े}} + V_{\text{शहरों}} + V_{\text{समस्या}} + R_{\text{आम}1} + V_{\text{इंसान}}$$
$$SDV_{\text{आम}2} = V_{\text{बड़े}} + V_{\text{शहरों}} + V_{\text{समस्या}} + R_{\text{आम}2} + V_{\text{इंसान}}$$
$$SDV_{\text{आम}3} = V_{\text{बड़े}} + V_{\text{शहरों}} + V_{\text{समस्या}} + R_{\text{आम}3} + V_{\text{इंसान}}$$
$$SDV_{\text{आम}4} = V_{\text{बड़े}} + V_{\text{शहरों}} + V_{\text{समस्या}} + R_{\text{आम}4} + V_{\text{इंसान}}$$
$$SDV_{\text{आम}5} = V_{\text{बड़े}} + V_{\text{शहरों}} + V_{\text{समस्या}} + R_{\text{आम}5} + V_{\text{इंसान}}$$
$$SDV_{\text{आम}6} = V_{\text{बड़े}} + V_{\text{शहरों}} + V_{\text{समस्या}} + R_{\text{आम}6} + V_{\text{इंसान}}$$
$$SDV_{\text{आम}7} = V_{\text{बड़े}} + V_{\text{शहरों}} + V_{\text{समस्या}} + R_{\text{आम}7} + V_{\text{इंसान}}$$
$$SDV_{\text{आम}8} = V_{\text{बड़े}} + V_{\text{शहरों}} + V_{\text{समस्या}} + R_{\text{आम}8} + V_{\text{इंसान}}$$

Out of these 8 Sentence Decode Vectors, the closest vector to the Test Sentence Vector should be used for the translation of the ambiguous word.

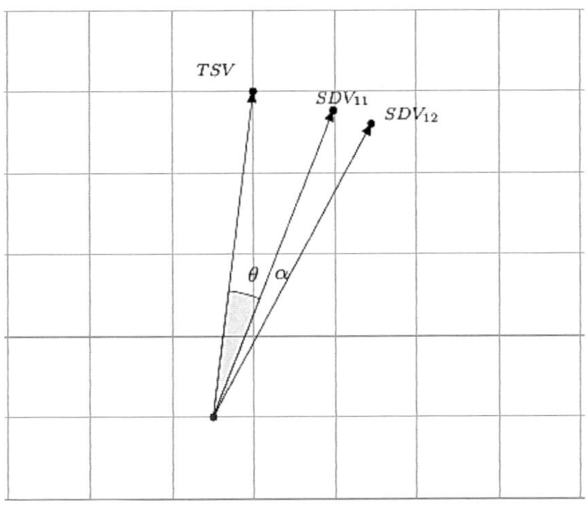

Figure 5. Cosine Distance Between Test Sentence Vector and Sentence Decode Vectors

To check the similarity, the cosine distance is calculated between Test Sentence Vector (TSV) and each of Sentence Decode Vector (SDV). To calculate similarity, we need to compute the TSV vector, which can be done like this:

$$TSV = V_{\text{बड़े}} + V_{\text{शहरों}} + V_{\text{समस्या}} + V_{\text{आम}} + V_{\text{इंसान}}$$

All the vectors for TSV are taken directly from the vector repository. Now the similarity between TSV and SDV can be calculated by this equation:

$$sim(TSV, SDV_i) = \frac{TSV.SDV_i}{||TSV||||SDV_i||}$$

Where $||TSV||$ denotes Euclidean norm of TSV and $||SDV_i||$ refers to Euclidean norm of SDV_i

If two vectors are identical than the cosine value will be zero and orthogonal relation between vectors indicates there is no match. The value of angle reaching to zero (or cosine value reaching to one) specifies that the meanings of two words are quite close to each other. The similarity scores are mentioned in the following Table-4.

IOP Conf. Series: Materials Science and Engineering 1131 (2021) 012022 doi:10.1088/1757-899X/1131/1/012022

Table 4. Cosine Similarity Score

Gloss/ Sense Id	Senses	Similarity Equation	Similarity Score	Rank
1	एक फल जो खाया या चूसा जाता है	sim(TSV, SDV_1) =	88.72	4
2	जिसमें कोई विशेषता न हो या अच्छे से कुछ हल्के दरज़े का	sim(TSV, SDV_2) =	91.29	2
3	गर्म देशों में पाया जाने वाला एक बड़ा, सदाबहार पेड़ जिसके रसीले फल खाए या चूसे जाते हैं	sim(TSV, SDV_3) =	89.84	3
4	प्रायः सभी व्यक्तियों, अवसरों, अवस्थाओं आदि में पाया जाने वाला या उनसे संबंध रखने वाला	sim(TSV, SDV_4) =	93.48	1
5	जो पका हुआ न हो	sim(TSV, SDV_5) =	79.35	5
6	जो कम पका हुआ हो	sim(TSV, SDV_6) =	78.62	6
7	एक रोग जिसमें शरीर से चिकना, सफ़ेद, लसदार मल बार-बार निकलता है	sim(TSV, SDV_7) =	74.91	8
8	एक प्रकार का लसदार चिकना सफेद मल जो अन्न न पचने से उत्पन्न होता है	sim(TSV, SDV_8) =	75.54	7
9[a]	Blank	-	-	-

a. 9 Synset fetched from IndoWordNet for "आम" including 1 blank Synset.

Further, we tried our approach with other meanings of the same word – "आम" with two more Test Sentences:

Test Sentence 2:

मुझे फलों में आम बड़े पसंद हैं |

Test Sentence 3:

पाचनतंत्र दुरुस्त न होने से भोजन अधपचा रहने की समस्या को आम कहते हैं।

The Test Sentence-2 has two words common with Test Sentence-1 i.e. "बड़े" and "आम". Similarly, the Test Sentence-3 has "समस्या" and "आम" common with Test Sentence-1. Now, we need to create Context Vectors for Test Sentence 2 and Test Sentence 3:

CV_2 = (V(फलों) + V(बड़े) + V(पसंद))

CV3 = (V(पाचनतंत्र) + V(दुरुस्त) + V(भोजन) + V(अधपचा) + V(समस्या))

The sense vectors are the same because of the same ambiguous word. However, the Reference Vectors as well as Sentence Decode Vectors will be different for each sentence.

5. Results

The closest senses of the ambiguous word-"आम" computed by our approach for each of the three above-mentioned Test Sentences are shown in Table-5.

IOP Conf. Series: Materials Science and Engineering 1131 (2021) 012022 doi:10.1088/1757-899X/1131/1/012022

Table 5. Sense disambiguation of the word-"आम"

Sentence	Total Ambiguous Senses	Number of Close Senses	Most Close Sense
बड़े शहरों में तो यह समस्या आम इंसान की है |	8[a]	2	प्रायः सभी व्यक्तियों, अवसरों, अवस्थाओं आदि में पाया जाने वाला या उनसे संबंध रखने वाला
मुझे फलों में आम बड़े पसंद हैं |	8[a]	4	एक फल जो खाया या चूसा जाता है
पाचनतंत्र दुरुस्त न होने से भोजन अधपचा रहने की समस्या को आम कहते हैं ।	8[a]	2	एक प्रकार का लसदार चिकना सफेद मल जो अन्न न पचने से उत्पन्न होता है

[a] 8 Synset fetched from IndoWordNet for "आम" excluding 1 blank Synset.

Performance of our proposed approach is compared with a method designed by [17] and shown in Table-6 with setup of same tuning parameters.

Table 6. Impact of variation in tuning parameters

Model	Tuning Parameters			WSD Precision	
	Vector Dimension	*Window Size*	*Negative Sampling*	*Previous Approach [17]*	*Proposed Approach*
SkipGram	100	15	0	45.00	49.47
	150	10	5	47.00	51.90
	200	5	10	36.00	41.27
CBOW	100	15	0	41.00	44.90
	150	10	5	52.00	57.21
	200	5	10	46.00	49.44

The next table displays the performance comparison of Individual words.

Table 7. Performance Comparison

Ambiguous Word	Number of Senses		Precision	
	Mentioned and Used in the Previous Approach [17]	*As per Indowordnet & used in our proposed approach*	*Previous Approach [17]*	*Proposed Approach*
आम	4	9*	85.71	91.49
सोना	7	5	14.28	32.18
उत्तर	3	14	28.57	35.40
शाखा	4	6	71.42	71.36
रंग	5	5	100	84.11
मंगल	5	5	57.14	65.78
फल	9	9	85.71	89.5
अंग	5	5	71.42	74.96
दल	12	12	0	24.62

6. Conclusion and future work

This paper proposes a framework to resolve the challenge of ambiguous words using word vectors. The proposed adaptive approach has outperformed the previous approach [17] which is also using word vectors to resolve the ambiguity. The best performance is shown by CBOW model with vector dimensions of 150, window size as 10 and a negative sampling of 5 words. The proposed adaptive approach with the CBOW model and particularly these tuning parameters has shown more than 5% of jump in performance from the previous approach.

The lemmatization of words is not performed before training the model which can further enhance the result. Another challenge is to disambiguate words which are used in idioms. Some of these words have completely different interpretations as to their role in idioms which are completely different from their present context.

References

[1] Ide, Nancy, and Jean Véronis. "Introduction to the special issue on word sense disambiguation: the state of the art." Computational linguistics 24, no. 1 (1998): 1-40.

[2] Agirre, Eneko, and Mark Stevenson. "Knowledge sources for WSD." In Word Sense Disambiguation, pp. 217-251. Springer, Dordrecht, 2007.

[3] Navigli, Roberto. "Word sense disambiguation: A survey." ACM computing surveys (CSUR) 41, no. 2 (2009): 1-69.

[4] Zhong, Zhi, and Hwee Tou Ng. "It makes sense: A wide-coverage word sense disambiguation system for free text." In Proceedings of the ACL 2010 system demonstrations, pp. 78-83. 2010.

[5] Taghipour, Kaveh, and Hwee Tou Ng. "Semi-supervised word sense disambiguation using word embeddings in general and specific domains." In Proceedings of the 2015 conference of the North American chapter of the association for computational linguistics: human language technologies, pp. 314-323. 2015.

[6] Rothe, Sascha, and Hinrich Schütze. "Autoextend: Extending word embeddings to embeddings for synsets and lexemes." arXiv:1507.01127 (2015).

[7] Chen, Xinxiong, Zhiyuan Liu, and Maosong Sun. "A unified model for word sense representation and disambiguation." In Proceedings of the 2014 Conference on Empirical Methods in Natural Language Processing (EMNLP), pp. 1025-1035. 2014.

[8] Mahrishi M., Morwal S. (2020) "Index Point Detection and Semantic Indexing of Videos—A Comparative Review". In: Pant M., Kumar Sharma T., Arya R., Sahana B., Zolfagharinia H. (eds) Soft Computing: Theories and Applications. Advances in Intelligent Systems and Computing, vol 1154. Springer, Singapore.

[9] Mishra, Neetu, Shashi Yadav, and Tanveer J. Siddiqui. "An unsupervised approach to Hindi word sense disambiguation." In Proceedings of the first international conference on intelligent human computer interaction, pp. 327-335. Springer, New Delhi, 2009.

[10] Singh, Satyendr, and Tanveer J. Siddiqui. "Role of semantic relations in Hindi word sense disambiguation." Procedia Computer Science 46 (2015): 240-248.

[11] Agarwal, Madhavi, and Jyoti Bajpai. "Correlation based word sense disambiguation." In 2014 Seventh International Conference on Contemporary Computing (IC3), pp. 382-386. IEEE, 2014.

[12] Patel, Nirali, Bhargesh Patel, Rajvi Parikh, and Brijesh Bhatt. "Hierarchical clustering technique for word sense disambiguation using Hindi WordNet." In 2015 5th Nirma University International Conference on Engineering (NUiCONE), pp. 1-5. IEEE, 2015.

[13] Tayal, Devendra K., Leena Ahuja, and Shreya Chhabra. "Word sense disambiguation in Hindi language using hyperspace analogue to language and fuzzy c-means clustering." In Proceedings of the 12th International Conference on Natural Language Processing, pp. 49-58. 2015.

[14] Singh, Satyendr, and Tanveer J. Siddiqui. "Sense Annotated Hindi Corpus." In 2016 International Conference on Asian Language Processing (IALP), pp. 22-25. IEEE, 2016.

[15] Athaiya, Anidhya, Deepa Modi, and Gunjan Pareek. "A Genetic Algorithm Based Approach for Hindi Word Sense Disambiguation." In 2018 3rd International Conference on Communication and Electronics Systems (ICCES), pp. 11-14. IEEE, 2018.

[16] Sharma, P., and N. Joshi. "Knowledge-Based Method for Word Sense Disambiguation by Using Hindi WordNet." Engineering, Technology & Applied Science Research 9, no. 2 (2019): 3985-3989.
[17] Kumari, Archana, and D. K. Lobiyal. "Word2vec's Distributed Word Representation for Hindi Word Sense Disambiguation." In International Conference on Distributed Computing and Internet Technology, pp. 325-335. Springer, Cham, 2020.
[18] Rumelhart, D.E., Hintont, G.E., Williams, R.J.: Learning representations by backpropagating errors. Nature 323, 533–536 (1986)
[19] Bengio, Y., Schwenk, H., Senécal, J.S., Morin, F., Gauvain, J.L.: Neural probabilistic language models. In: Holmes, D.E., Jain, L.C. (eds.) Innovations in Machine Learning. STUDFUZZ, vol. 194, pp. 137–186. Springer, Heidelberg. 2006.
[20] R. Collobert and J. Weston: A Unified Architecture for Natural Language Processing: Deep Neural Networks with Multitask Learning. Architecture, vol. 20, no. 1, pp. 160 – 167, 2008
[21] T. Mikolov, G. Corrado, K. Chen, and J. Dean: Efficient Estimation of Word Representations in Vector Space. Proc. Int. Conf. Learn. Represent. (ICLR 2013), pp. 112, 2013
[22] T. Mikolov, K. Chen, G. Corrado, and J. Dean: Distributed Representations of Words and Phrases and their Compositionality. Nips, pp. 1 – 9, 2013
[23] A. Joulin, E. Grave, P. Bojanowski, and T. Mikolov: Bag of tricks for efficient text classification. arXiv Prepr. arXiv1607.01759 (2016)
[24] Soni, Vimal Kumar, Dinesh Gopalani, and M. C. Govil. "Resurgence of Deep Learning: Genesis of Word Embedding." In Smart Innovations in Communication and Computational Sciences, pp. 129-139. Springer, Singapore, 2019.
[25] Siddiqi, Sifatullah, and Aditi Sharan. "Construction of a generic stopwords list for Hindi language without corpus statistics." International Journal of Advanced Computer Research 8, no. 34 (2018): 35-40.

AUTHOR INDEX

Institute of Physics
Dirac House, Temple Back
Bristol BS1 6BE UK

ISSN: 1757-8981
ISBN 978-1-7138-3409-0

CURRAN ASSOCIATES INC.
proceedings
.com